Dampness in Buildings

D1330662

To all those who live in houses and to those who look at houses on behalf of those who live in them – this book is dedicated.

Dampness in Buildings

Diagnosis, Treatment, Instruments

Second Edition

T. A. Oxley
Formerly Hon. Professor of Biodeterioration,
Aston University, Birmingham

E. G. Gobert
Hon. Life President, Protimeter plc, Marlow

BUTTERWORTH
HEINEMANN

Butterworth-Heinemann Ltd
Linacre House, Jordan Hill, Oxford OX2 8DP

$\mathcal{R}$ A member of the Reed Elsevier group

OXFORD   LONDON   BOSTON
MUNICH   NEW DELHI   SINGAPORE   SYDNEY
TOKYO   TORONTO   WELLINGTON

First published 1983
Reprinted 1985, 1987 (revised), 1989 (revised)
Second edition 1994

British Library Cataloguing in Publication Data
Oxley, T. A.
   Dampness in buildings – 2nd ed.
   1. Dampness in buildings
   I. Title II. Gobert, E. G.
   693.8'93   TH9031

ISBN 0 7506 2059 5

Library of Congress Cataloguing in Publication Data
Oxley, T. A. (Thomas Alan)
   Dampness in buildings
   Includes index.
   ISBN 0 7506 2059 5
   1. Dampness in buildings I. Gobert, E. G. II. Title
   TH9031.094   1994   693.8'93   83–17152

Printed and bound in Great Britain by
Biddles Ltd, Guildford and King's Lynn

Contents

Preface

This book deals with dampness problems in buildings and their solution.

Buildings

Stage 1. Early Warning

Every survey of a house includes a survey for dampness. Since 1956 surveyors have been using a Protimeter moisture meter to avoid missing a dampness problem; for dampness can be hazardous long before it can be detected by the human senses.

The difference of 3 or 4 per cent between what is air-dry in wood and in buildings and what is dangerously damp is too small to be 'detected by the human eye or by placing of hands against walls or the like. . .' said Judge Newey QC in *Fryer v Bunney* (Official Referee's Court, 10 November 1981).

He went on '. . . but I am abundantly satisfied that it (the dampness) could have been detected by the use of a Protimeter (moisture meter). . .'.

Stage 2. A Complete Diagnosis

It would be foolish to carry out a survey without a moisture meter – but it would be no less foolish to rely on a moisture meter on its own for a complete diagnosis. Far too often condensation is mistaken for rising dampness, and the wrong (expensive) cure is prescribed; and occasionally contamination by soil salts is put forward as a reason for an incorrect reading. There are now tools available which make an incorrect diagnosis a thing of the past.

Painting

Using instruments now readily available it is possible to make sure that a surface is dry enough to paint and that dew (condensation) is not about to form on a metal surface.

Floors

A new sand/cement screed is often the slowest part of a building to dry out. It is important, therefore, prior to laying a floor-covering, to know the moisture level of a solid floor. British Standards 5325: 1983 and 8201/3: 1987 describe two methods: the 'conductivity' test; and the 'hygrometer' test. There is at least one instrument available capable of carrying out both tests.

Structural concrete

This can be tested for moisture using a concrete probe, an accessory of a chilled-mirror dew-point meter which is one of the most modern and accurate humidity measurement instruments available anywhere.

Wood

An oven-test can be misleading. A moisture meter for wood is better than the oven as it shows the distribution of moisture inside the wood. In this case the moisture meter beats the oven.

Note: As dries the air, so shrinks the wood (in storage). For the lumber-man an inexpensive hygrometer is also indispensable.

Sources of information

We should like to thank the several organizations which have allowed us to reproduce extracts from their publications.

Extracts from British Standards are reproduced by permission of the British Standards Institution, 2 Park Street, London, W1A 2BS, from whom complete copies of the standards can be obtained.

Extracts from Building Research Establishment (BRE) Digests Nos. 110, 245 and 270, are reproduced by permission of the Controller of HM Stationery Office.

Photographs *Figures 9.1* to *9.6* are Crown Copyright and reproduced by permission of BRE Princes Risborough Laboratory.

Figures 7.2A, B and C were prepared by Professor Oxley for the *Building Trades Journal* and are reproduced by permission of that journal.

Protimeter is a registered trade mark of Protimeter plc.

Our special thanks go to our colleagues Gerald Gobert, Dr Colin Kyte and Fil Dadachanji who will find many of their thoughts incorporated in these pages.

Introduction

A word for the professional

The purpose of this book is to provide a better understanding of dampness problems in buildings and to offer advice on diagnosis and cures; it is not to provide any data for the design of buildings.

In writing it we initially thought of addressing it only to you, the professional surveyor, architect, environmental health officer and builder. We well remember the enthusiasm with which the first Protimeter moisture meter was received by you in 1956. Many of you have since gone on record as saying that you do not know how you were able to carry out an efficient survey of a house without using a 'Protimeter'; and many more of you are proving it daily by your continued use of this type of meter. The collaboration between you, the professional and Protimeter plc has been a fruitful and happy one lasting over more than a quarter of a century, and is still continuing.

Moisture problems have always been the bane of your lives. And if anything, they are even worse today than they were in 1956. Condensation has become a much more common problem particularly as it can so often be confused with rising damp.

Indeed, we have good reason to believe that only about one third of all dampness problems are due to rising damp. The other two thirds are condensation and other forms of moisture ingress:

First there is the survey carried out by the Building Research Establishment Scottish Laboratory between November 1979 and March 1980, the results of which were published by HMSO in 1982 under the title 'Dampness: one week's complaints in five local authorities in England and Wales'. One of the significant conclusions of the survey is that 66 per cent of the identified dampness complaints were caused by condensation.

Secondly, in the Protimeter Laboratories specimens of wallpaper and plaster are received almost daily from surveyors and local authorities for chemical analysis for the presence or absence of certain nitrate and chloride salts, which are the typical by-products of rising dampness (see Chapter 4).

Although the specimens obviously come from walls where the surveyor or environmental health officer suspects the possibility of rising damp, yet salts are consistently found from year to year to be present in only about one third of all specimens tested.

This is proof, if proof is needed, of how very difficult it often is to diagnose the true cause of dampness; and whilst there are, of course, many conscientious remedial treatment firms, there are inevitably some which are keen to carry out a profitable cure for rising damp irrespective of the true fault; and if you happen not to agree with their prognosis, they produce ingenious gadgets and arguments in support of their case.

This book will help you where appropriate to prove them wrong.

Having said this we hope you will not mind if we address this book also

To the practical and concerned house owner

If you are concerned about the condition of your home, about your health and comfort, about possible decay, damage, and loss in value, you should read this book – or most of it. It will help you to carry out an inspection of your house regularly so that you know the extent of any problem. Every house has a few damp patches. Usually they don't signify greatly; often they appear in winter and disappear during the following summer without any damage having been done.

You can't see these damp patches. If a wall or window frame is so wet that you can see and feel that it is wet, it is time to do something urgent about it, and it may be too late for a DIY job. The damp patches which come and go (which needn't worry you if they *do* go again) can only be detected with an electrical moisture meter. But if they remain throughout the year, they will be getting gradually worse because there must be a source of water continually acting, and decay of wood will inevitably follow. You must do something about these even though there may be nothing obvious at first.

If you inspect your house at least twice a year (about one or two hours' work, preferably at the end of the winter and again at the end of the summer) you will know where there is any damp and whether it is a matter for action or not. By following the guidance supplied with the meter and using, if necessary, the wall salts analysis service provided by Protimeter plc, you should find it possible to diagnose the cause of any dampness you find. You will be able to tell whether or not it is a matter for concern and whether it is likely to persist or get worse. You will be able to deal with the problem before it becomes serious and expensive.

Do you need a specialist firm?

You may be able to cure the trouble yourself or it may be better to call in a specialist firm. But it is much better to call in a specialist and tell him what needs doing, than to invite a 'free' survey which must be paid for eventually out of the profit on the job; in the long run, no survey can be truly free.

We urge you to help yourself in matters of dampness in the home. Know and understand what is the true condition of your house; diagnose and cure any dampness before it can cause you any trouble. Avoid being 'sold a pup' in the house you are buying. Above all, avoid putting yourself in the hands of firms which have a vested interest. Use specialists only to do what really needs to be done.

You can't do this without a moisture meter. All professional surveyors use them; by far the greatest number use Protimeter instruments, sophisticated and highly sensitive tools appropriate to their professional skills. But for detection and diagnosis of dampness before you can feel it or see it, and for finding the source of any obvious dampness which you can see, there are one or two small meters available to you which are comparatively easy to use. They are sufficiently sensitive and extremely portable. You can use them on top of a ladder, in the loft, or in the space under the floor.

When to call in a surveyor

In one respect you, the householder, have an advantage over the professional, because you can inspect regularly. You can keep odd damp patches under observation to see whether they are growing or declining and so have the benefit of monitoring changes which the professional surveyor on a single visit cannot do. You need not call in a professional surveyor for simple problems, but you should do so, as one calls in a doctor, if a problem gets beyond you. But, even if you find you need outside help, at least by carrying out regular instrumental inspections you may be able to prevent a small problem becoming a large and expensive one.

For the house owner, detailed guidance on how to conduct a survey for dampness in your own home is given in the Appendix of this book.

Remember that your home is a big and worthwhile investment. Do look after it.

1

The dampness problem

Those who have responsibility for buildings, and those who use them, are more conscious of dampness now than in the past. If they are surveyors, they have onerous responsibilities which demand that they overlook no sign of dampness, and they are expected to advise their clients on the extent, severity, and future implications of what they find. If they are building or housing managers, or enviromental managers, they will be left in no doubt by the occupants if dampness is among the defects that they suffer, and they will also be expected to prescribe the cure. But a cure is only possible on the basis of a correct diagnosis. This book, therefore, is to help the reader towards a correct diagnosis.

The great majority of surveyors use electrical meters based on the measurement of conductance for detection and evaluation of dampness; their job, insofar as it concerns building dampness, would be impossible without them. Many building, housing management, and environmental health departments also find these instruments essential. But dampness is usually not a simple problem and hence correct diagnosis is sometimes very difficult. Electrical instruments will not, themselves, diagnose the cause of dampness. They do, however, provide indications of great value, including the quantification of dampness which might otherwise not yet be detectable.

An electrical moisture meter for buildings is therefore part of a system for diagnosis, of which the other part is the knowledge and experience of the operator. It is the aim of this book to provide the knowledge and to show how a range of conductance type moisture meters, together with their accessories, and an analytical service, can provide the evidence. The remaining factor, the experience, we cannot supply. But experience will be gained more quickly if interpretations are based on understanding, which we hope this book will provide.

All dampness is water out of place, but it is convenient to classify its different manifestations by their source, and the routes by which the unwanted water enters the inhabited areas. A primary distinction is between water which enters as a liquid, and water which is condensed from the atmosphere. Until the 1960s the latter was almost unknown within buildings, but with changes in building design and in living styles, condensation has become a major problem, causing about two thirds of the complaints of dampness in houses received by housing authorities. It is this huge increase, and the difficulty of diagnosis which it often presents, which is partly responsible for the greater awareness of dampness among both professionals and laymen. Another factor is the great amount of rehabilitation of older buildings, large and small, which is now undertaken. This has often revealed dampness which was tolerated, or regarded as inevitable, by our forebears; in addition 'modernization' does much to increase condensation by total elimination of draughts and modern decoration makes it more obvious.

Awareness of dampness has also been stimulated by the rise of a service industry of 'specialist' firms devoted to curing it. This is an industry largely directed towards curing rising damp. It is generally based on the valuable and relatively cheap process of injecting water repellent materials into the lower parts of walls, thereby exploiting a principle explained later in this book for preventing the rise of water by capillarity. This is a competitive industry which uses a lot of publicity; it has spread quite widely the impression that rising damp is the main cause, or at least a very frequent cause, of dampness in buildings. In fact rising damp is relatively uncommon cause of dampness in buildings.

The specialist damp-proofing industry includes a majority of responsible firms. Many of them are members of the British Chemical Dampcourse Association and similar organizations in other countries. These firms work to the excellent codes of practice published by their Associations. However, the standard of others is not universally high. It is necessary, especially, to look critically at the survey reports produced by some who are surveyors in name only. It is one function of this book to aid such critical evaluation. It is hoped that critical customer appraisal will bring to an end the practice of attempting to cure condensation, penetrating damp, or even plumbing leaks, by irrelevant treatments for 'rising damp'.

Decay of wood is one of the most serious consequences of dampness. It is therefore a considerable advantage of the conductance type of electrical moisture meters that they give direct measurements of water in wood. With the aid of a deep penetrating hammer electrode, the instruments especially designed for timber can show whether or not there is serious dampness within thicker pieces which is not apparent at the surface. This is important when structures are drying out after flooding or saturation during fire fighting.

If, as sometimes happens, it is impossible or uneconomic to cure a source of dampness completely, it is important to treat wood with preservative, or to use wood which has already been so treated, for replacement. Wood preservative treatments, or methods of application, are not all equally effective;

some may be little better than cosmetic. Some other products are designed to be diluted on site, and if this is poorly controlled the results may be correspondingly poor. The best safeguard is to use firms who are members of the British Wood Preserving Association, the American Wood Preservers Association, or similar responsible organizations, and to use products made by such firms. The Associations' codes of practice and lists of approved products for use in remedial or pretreatment preservation are a valuable guide.

Figure 1.1 Surveying for dampness in the house

Although this book is primarily concerned with diagnosis, we believe that it is useful to give some guidance on techniques for the cure of dampness. Therefore, each of the sections of this book on the various types of dampness problem either contains, or is followed by, an 'answer' outlining the principles and methods appropriate to each. These are not intended to be comprehensive, but we hope they will guard against the use of irrelevant or inappropriate measures.

Beware of the man who does not use a moisture meter because he claims it is 'misleading'. It is more correct to say that the meter is 'revealing', because without it, much significant dampness would pass unnoticed. Without the graduations in dampness which a meter gives, the further information given by its accessories and the associated laboratory service, it would often be impossible to diagnose the true cause, especially in the early stages.

2

What dampness is

Dampness in buildings, if left unattended, can lead to structural deterioration; it will result in the decay of wood, it will spoil decorations and, by encouraging development of moulds and mites, can be dangerous to health.

If an absorbent material such as a piece of wood, paper, cloth or brick is placed in a very damp atmosphere (high relative humidity) it will absorb water and therefore increase in moisture content. Conversely, in a dry atmosphere (low relative humidity) it will lose water and its moisture content will fall. At any intermediate relative humidity of the air the material will either gain or lose moisture, depending on whether it was dry or wet to begin with, and finally it will settle to a certain moisture content which will remain constant so long as the humidity of the air does not change. The material is then 'in equilibrium' with the air.

If the humidity changes, the moisture content of the material will gradually change with it. So for every relative humidity there is a definite moisture content at which any piece of absorbent material will be in equilibrium. But this figure is different for almost every kind of building material, and even for almost every piece; certainly it varies from brick to brick and from one kind of mortar to another. But for wood, although there are differences between different varieties, and

Box 1

How moisture is measured in a laboratory

The basic measure of moisture content is made by oven drying. A sample is weighed, dried, and weighed again. The loss in weight is assumed to be water and this is expressed as a percentage of the final oven dry weight.

Suppose a piece of brick is to be tested for its moisture content. It must, of course, have been wrapped in foil or several layers of polythene, otherwise it will have lost or gained moisture since it was removed from the structure. If large enough, it should be broken into two or three so that the determination can be carried out in duplicate or triplicate. Assume that it is broken into three pieces, dust is rejected, and the pieces are weighed.

It is specified that the sample must be dried to 'constant weight' at a temperature of 105 °C. It is therefore dried in a thermostatically controlled oven, probably for two hours in the first instance, then for a further four hours, and then a further 16 hours:

	Sample 1	Sample 2	Sample 3
Original weight/g	3.721	10.086	2.820
Weight after 1st drying/g	3.629	9.788	2.735
Weight after 2nd drying/g	3.582	9.710	2.719
Weight after 3rd drying/g	3.580	9.705	2.712
Calculated moisture content/%	3.94	3.93	3.98
Average = 3.95 per cent reported as 4.0 per cent			

In future the laboratory would always dry material of this type for about 18 hours having found that 6 hours is not sufficient. The laboratory would be careful to use a lower temperature for samples containing gypsum plaster because this material decomposes at high temperatures and will lose 16 or 17 per cent of its weight. This is chemically combined water (see Chapter 2) and not dampness. If a moisture as high as 16 per cent is reported by a laboratory you will know that it has been overheated. Even saturated gypsum plasters will not hold more than a few per cent by weight of water.

between hardwoods and softwoods, there is not nearly as great a difference as there is between other building materials. This is why moisture meters usually give a moisture content scale for wood, but do not attempt it for other materials. The accompanying graph (*Figure 2.1*) shows the average humidity/moisture content relationship for typical softwoods used in buildings.

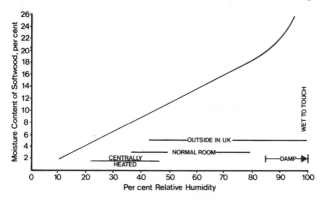

Figure 2.1 Wood moisture content and air relative humidity

The graph shows the approximate relationship between the relative humidity of air and the moisture content of wood. This applies to typical softwoods used in building; the curve for some of the heavier hardwoods would be different. The horizontal lines show the typical range of humidities met in various circumstances. Wood kept in these environments will gradually come into equilibrium at the moisture content levels indicated on the vertical scale. This shows why wood becomes very dry (water content 4–8 per cent) and often shrinks and cracks in centrally heated rooms, but will become much damper, though not dangerously damp (up to 16–18 per cent) in a normal room. Wood exposed to outside air continuously changes in water content over a very wide range. Although it becomes very wet in wet weather it does not decay so long as it has the opportunity to dry out in dry weather

The percentage moisture figures are derived from oven drying tests and are based on this universally accepted formula:

$$\frac{\text{Wet weight of the material} - \text{dry weight of the material}}{\text{Dry weight of the material}} \times 100$$

In the test the sample is very accurately weighed (wet weight), and it is dried in a ventilated hot air oven especially designed for the purpose at an accurately controlled temperature until it reaches a constant weight, that is, until it loses no more moisture. It is then allowed to cool in specially dried air. When cool it is again accurately weighted (dry weight). The loss in weight due to evaporation of water is expressed as a percentage of the final weight as shown above.

It follows that a heavy material has a much lower percentage moisture content than a light material which has the same amount of water in it!

Graphs could also be drawn for every other building material, but the materials are so immensely variable that such graphs would probably be different for every brick, every sample of mortar, plaster, concrete or wallboard, and all would be very different from wood. If several different materials are built into the same wall the effect of this will become obvious. For example: Wood battens are set into brickwork covered with wallboard on one side, and plaster on the other; they will exchange moisture with each other, and with the air, until all have come into equilibrium. Suppose the atmosphere is at a relative humidity averaging 50 per cent; it will be seen that the moisture content of the wood is just under 11 per cent. But the bricks may vary between, perhaps, 1½ and 2½ per cent, the plaster probably less than 1 per cent and the wallboard perhaps 9 or 10 per cent. This is the normal condition of a perfectly normal wall.

Now if this wall becomes damp, all the materials will share in the dampness until they are again in equilibrium; their moisture contents will be higher, but will continue to be widely different. If the air has not changed they will begin to lose moisture to it as they dry out. But if, when the wall became damp the air also became damper and happened to remain in equilibrium with the wall, its various components would not dry out, and they would remain at their higher (and widely different) moisture contents.

Obviously it is meaningless to quote a moisture content for such a diverse structure and to say that such and such a moisture is 'dry' or 'wet' unless you know the characteristics of each of the components, which in practice you never will (see Chapter 7). The single common component is the relative humidity of the air. If you can say that a particular structure is in equilibrium with a particular relative humidity you have made a very useful statement about the dampness of the structure, because *it is the relative humidity which determines whether or not moulds will grow, decay fungi develop in wood, or decorations be damaged.* It is useless, from this point of view, simply to quote the moisture content (unless you are talking about wood).

Now it is possible to answer the question, 'what dampness is'. It is usual to say that a material is damp if it is wetter than 'air-dry' as defined below.

We said earlier that at the relative humidity of 50 per cent, wood will contain about 11 per cent moisture, that is about one tenth of its weight is water. Yet it is regarded as 'dry' although 'air-dry' is a more accurate description. It follows that 'air-dry' means 'in equilibrium with a "normal" atmosphere'; although 'normal' atmospheres do vary considerably, from as low as 30 per cent relative humidity in a well-ventilated centrally heated office to 70 per cent relative humidity in a busy classroom. Air-dry wood defined in this way (as can be seen from the graph) has a moisture content of between about 6 and 16 per cent.

In this book we take 'air-dry' to mean the condition of a material in an ordinary indoor, inhabited environment with a relative humidity not exceeding about 70 per cent.

A definition of 'damp'

From the above discussion follows the definition of 'damp'. The serious complaints about dampness relate to development of moulds, spoilage of decorations, decay of wood and wood-based materials, appearance of mites (minute eight-legged creatures related to spiders) and possible adverse effects on health. All these have a biological origin. Each of these requires dampness to develop and they all have a similar limit of dryness below which they cannot live or multiply. It is reasonable to take this limit as the line between dry and damp in buildings. It is not a precise line because there is a range, between about 75 and 85 per cent relative humidity, in which the offending organisms can develop, but very slowly and without causing much trouble. For wood (see Figure 2.1) this is between 18 and 20 per cent moisture content. Above 85 per cent relative humidity, moulds, decay fungi, and mites can develop quite quickly, the rate becoming faster and more troublesome at higher levels.

We can therefore say that 'damp' is an atmosphere wetter than 85 per cent relative humidity; and a material is 'damp' if it is in equilibrium with this humidity.

Wet wall, dry air

The air will not necessarily be at this high humidity if the dampness is caused by water from a source other than high air humidity such as rising or penetrating damp or a plumbing leak. If the wall is damp (wetter than air-dry) it will be losing water to the air all the time. But while it is still wet, the air in a layer close to its surface, and in any cracks, will be in equilibrium with it regardless of the dryness of the general air of the room. So the moulds, fungi and mites which are the adverse consequences of 'dampness', being very small, are affected by the equilibrium humidity of the wall, not by the actual humidity around it. In due course, when the wall dries, as it will if the air continues to be dry and there is no continuing source of water, the offending organisms will die, being dried up. But if there is a continuing source of water, such as a plumbing leak, water penetration from the outside, or rising damp, the wall will not dry even though the air in the room is dry. So moulds, fungi and mites will be able to grow on and within a wet wall even in a dry room.

Measuring dampness

Since moisture content is such a poor measure of the dampness of a wall which is wetter than air-dry, what alternative is there? The theoretical ideal is to cover the suspect damp area with a waterproof tent of polythene, or foil, or with a box, and have a humidity measuring device under it. Water evaporating from the wall, into the small amount of air trapped in the tent or box, will raise its relative humidity until it is in equilibrium. Then, by measuring the relative humidity it is possible to say exactly how damp the wall is, regardless of the humidity in the room as a whole.

Obviously this is an impossibly laborious process for surveying a building, for it would take several hours at each point. However it is the recommended method (British Standard BS 5325:1983) for determining whether or not a concrete slab is dry enough for a moisture-sensitive floor to be laid.

Fortunately this is not the only method. The relative readings of an electrical moisture meter (especially of the conductance type) measure only the free water in a material; therefore they closely indicate the relative dampness of different materials. Although they do not measure relative humidity, their indications are a fairly close representation of it. Thus a high reading on such a meter (in the absence of contaminating salts or carbonaceous materials) indicates a damp condition of

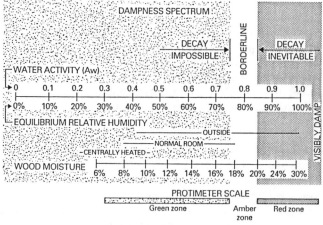

All Protimeter moisture meters for building surveying have a colour-coded scale to replicate these three levels of moisture found in buildings.

Figure 2.2 A 'dampness spectrum'. The relations between several different ways of expressing the significance of water in material are illustrated in the diagram which can be called a 'dampness spectrum'. It is drawn on a regular scale of relative humidity (RH) from zero to 100 per cent and exactly corresponding to this is a scale of 'water activity' (Aw) from zero to unity which is a very useful measure of wetness and dryness much used in industry.

Wood moisture content (average soft wood) is shown and the correspondence between moisture content and relative humidity is easily seen.

The 'visibly damp' indication is very approximate.

approximately equal significance in wood, brick, plaster or wallboard, regardless of their very different moisture contents. Therefore it is possible to mark on the scale of an electrical meter indications of 'safe', 'intermediate' and 'danger' which correspond reasonably well with the humidity equilibria of most non-metallic or non-carbonaceous materials on which they may be used. Some well-known instruments do this by a colour code: green indicates the 'safe' condition, corresponding to an 'air-dry' condition in an ordinary indoor, inhabited environment. Red indicates a humidity equilibrium in excess of about 85 per cent and a hatched or amber area indicates the region between.

Yet it is important to note that materials do not become visibly damp, and do not feel damp to the touch, at 85 per cent humidity equilibrium. Wood, for example, does not feel damp below 30 per cent moisture content (i.e. around 97 or 98 per cent relative humidity). Thus dampness is hazardous long before it can be detected by the unaided senses. This is why it is so essential to use a moisture meter when surveying for damp, and making judgements about its severity (*see also* Chapter 7; measurement of moisture on site is described in Chapter 3)

How water is held in building materials

There are three ways in which moisture is held in building materials:

(1) *Chemically combined (or 'bound') water*
    This is part of the water mixed with materials such as concrete and plaster during building construction. The amount of water used in construction of an average sized house can be as much as 4000 kg (4 tonnes) and the drying period may be as long as a year. During this period much of the moisture evaporates into the internal air of the building and may cause condensation, particularly where ventilation is poor and the heating intermittent. Once the drying out period is over, a quantity of water remains chemically combined in the material, which does *not* contribute to dampness problems. Water, when it is chemically combined in cement, gypsum, or other setting material is not water at all, it is permanently part of the set material.

(2) *Sorbed water*

As explained in previous pages most materials can take up water directly from the air, the amount of water absorbed depending upon the ambient humidity. Over a fairly narrow range most materials are changing in moisture content all the time in response to changes in the air humidity, but direct sorption of water from the air by uncontaminated materials does not give a dampness problem in buildings unless the ambient air humidity is persistently very high (*see Figure 2.2*). This is most common when condensation has become a problem. However, when materials have been contaminated with hygroscopic inorganic salts, an excessive amount of water may be absorbed directly from relatively dry air, and the material (usually plaster or paper) becomes visibly damp.

(3) *Capillary water*

Nearly all building materials have a porous (or capillary) structure, and if these pores (or capillaries) are filled with water, a serious dampness problem can result. It is when water is in capillary form that it can move through a material, rising from the foundations (rising damp) coming through walls (penetrating damp) or soaking into a wall subject to persistent condensation.

To sum up

Because significant dampness cannot be detected by the unaided senses, buildings ought to be regularly surveyed for damp by use of a meter. Following the instructions (*see Chapter 7*) it is possible to detect and pinpoint accurately damp spots or areas or, alternatively, to give a house a clean bill of health. But it is necessary to exercise judgement. When dampness is found it is necessary, first to satisfy yourself that you know the reason why the material is damp, secondly to decide whether it is transient, or likely to be persistent, and thirdly to take appropriate curative action if it is likely to persist. Over-hasty reaction to a single high moisture reading is not justified. Diagnosis of the cause of dampness is often not straightforward. It requires a professional, or well-instructed, approach and it requires tools.

The moisture meter is to the engineer, the architect and the surveyor what the stethoscope is to the doctor; it is a tool giving indications which cannot be gained from the unaided senses, but it requires understanding for correct diagnosis.

3

Is there a dampness problem?

Measuring dampness on site

We have shown that most dampness problems in buildings exist before they can be recognized by the human senses. It follows that the professional surveyor or the house owner carrying out his annual inspection must use one of the techniques available to make sure he finds all the areas where excess of moisture is present. Even if dampness is obvious, measurement will be needed (a) to see how damp it is and to assess future hazards and (b) outside the limits of obvious dampness to see how far the hazard extends and to pinpoint its source (see Chapter 7).

There are three basic techniques for measuring dampness in building materials:

(1) Sampling
    (a) Measurement of moisture content
    (b) Measurement of equilibrium relative humidity
(2) Water activity
(3) The use of moisture meters
    (a) Conductance type moisture meters (e.g. Protimeter)
    (b) Capacitance (dielectric) type moisture meters

Sampling

To obtain moisture content
If a moisture meter is not available, the usual alternative procedure is to take samples, a process which is somewhat destructive. It is carried out either by drilling and collecting the spoil removed by the drill, or by removing whole bricks, or large parts of them, with hammer and chisel. Obviously brick removal is a very drastic process taking much time and effort and causing immense disturbance and mess. It is only possible to do this at a few points so that it is impossible to map out dampness areas. Such a procedure is obviously not practicable for survey purposes.

Less disturbance is caused by drilling, and in principle it is possible to obtain quite a large number of samples and make some attempt to map out damp areas, with the advantage that dampness at various depths can also be determined. It is necessary to drill at a regular speed with a freshly sharpened bit to minimize heating, which would cause rapid loss of water from the small sample of brick dust, and to collect the sample immediately into an airtight container. Measurement of the moisture content in the sample thus obtained requires either full laboratory equipment (thermostatically controlled oven, balance sensitive to one milligram, drying tins and desiccator; *see* Chapter 2) or the use of an acetylene pressure type instrument. For the latter, at least three grams of brick dust is needed so that it is necessary to drill with quite a large bit. A standard amount of the sample is accurately weighed and immediately placed in a small pressure cylinder. A measured amount of calcium carbide powder is added. The lid is replaced and screwed down and the container is shaken to mix the carbide with the sample; the water in the sample reacts with the carbide producing acetylene gas. The pressure which builds up in the cylinder is therefore a measure of the amount of moisture in the sample. A pressure gauge calibrated in percentage moisture will give the moisture content of the sample based on the standard weight taken at the beginning. It is absolutely essential to keep the instrument in perfect condition, the washers regularly renewed and only fresh carbide powder used, or low readings will be obtained as a result of leaks. As acetylene gas is highly inflammable, smoking or the use of naked lights whilst using this instrument must be forbidden.

Although drilling is less destructive than removal of bricks, it causes damage to walls and decorations which would be unacceptable for survey purposes in most situations. In addition it is relatively slow; half a dozen readings obtained in an hour would be quick work. Obviously this is not a suitable method for survey work, and although it can be used to determine the moisture content of drilled samples taken from deep inside a wall irrespective of the presence or absence of hygroscopic salts, it cannot be used to obtain surface readings in walls nor can it be used to obtain moisture readings in wood. But, as has been shown in Chapter 2, moisture content is not a useful measure of the significance of moisture in a building material other than wood.

To obtain percentage relative humidity

The alternative of measuring the equilibrium relative humidity (ERH) – or the water activity (see below) – of a sample has the advantage of giving a result of direct significance independently of the nature of the material. But it requires a rather larger sample than can conveniently be obtained by drilling and it requires an accurate instrument for measurement of relative humidity of a kind which does not itself affect the atmosphere it is measuring. A direct reading dew point meter is ideal. Obviously this is not a method which can be used for survey work, but it has advantages for diagnosis which may, in difficult cases, justify its use. It is a useful measure of the significance of moisture in building materials. However, as it is expressed in terms of percentage ERH it is often confused with percentage moisture content. It is preferable therefore to use the concept of water activity.

Water activity

The concept of water activity (Aw) has been used in the food industry for many years for the measurement of water in materials. However, not many building surveyors are familiar with this term.

It refers to the availability of water in a solid (like wood) or a liquid (like a strong solution) to take part in chemical reactions or to support life.

Water activity is precisely analogous to relative humidity in an atmosphere, and expressed in a similar ratio except that relative humidity is expressed in percentages whereas water activity is expressed as a decimal fraction.

If, for example, a piece of wood is in equilibrium with a relative humidity of 85% the Aw of water in the wood is 0.85 which is a measure of the ability of moulds to obtain water inside the wood for destructive purposes. As is well documented, at this level of humidity the wood will rot. When the water activity is 0.7, the wood is safe.

A water activity test is carried out by taking a sample from the suspect area. This is placed in an airtight container with a water activity sensor in it. Under controlled temperature conditions, the material and the air inside the container come into moisture equilibrium. The equilibrium relative humidity, that is, the water activity is displayed.

An alternative and non-destructive method is to use

an inexpensive and disposable Protimeter Aw-patch. The patch is self-adhesive and consists of a humidity sensor between a waterproof and semi-permeable membrane. The patch is attached to the suspect wall surface. The air under the patch will come to equilibrium with the wall and the water activity can be measured with a Protimeter humidity measuring instrument.

Water activity tests are not affected by carbonaceous materials – and take account of contaminating salts which may be present. Therefore Aw test results are of undoubted and immediate practical value to any building inspector, architect, builder and housing manager.

The results show unequivocally to which zone of the Protimeter dampness spectrum (see page 18) a measurement belongs; that is, whether or not the area is safe from decay.

Moisture meters

The measurement of the equilibrium relative humidity or of the water activity of a material are excellent methods to establish the significance of moisture: they have an important role to play in the diagnosis of particularly difficult cases of dampness ingress, although they are not methods which can be used for surveying and for establishing quickly whether a problem exists or not. For surveying there is no practical alternative to the electrical moisture meter.

The rapid electrical methods for measurement of water in various materials operate on either the conductance principle, or the capacitance principle. Some instruments aim to combine both principles in various degrees. In evaluating the two systems it is helpful to understand, in outline, how they work.

Conductance

The dry substance of many materials (e.g. wood, paper, seeds, brick, concrete, etc.) forms an insulator, that is, it will not conduct electricity. However, such materials all absorb water very easily and when they do so they do begin to be able to conduct electricity. As water is added to the dry material, at first there is very little effect, but the addition of more water gradually increases its conductivity (i.e. its ability to pass a small current of electricity) in a regular way, so that by measuring conductivity to electricity it is possible to tell how much water has been added, up to a point beyond which further additions have little effect. This relation between the amount of water and the electrical conductivity of materials which are insulators when dry depends not on the conductivity of water itself, but on the conductivity of solutions which form, due to the water dissolving minute amounts of soluble materials. Many of these dissolved substances are ionizable, that is, when they dissolve each molecule splits into two parts, one with a positive charge and the other with a negative charge. When an electrical connection is made with the damp substance by means of metal electrodes, these charged particles (called ions) move through the water to the electrodes, the negative ions going to the positive electrode and the positive ions to the negative electrode. Here they give up their charges which results in a flow of electricity. Note that when a very small amount of water is added to a completely dry material, each water molecule is tightly adsorbed on to the dry surfaces. The molecules of water are held very firmly in this position. They are not free to move or behave like water in any way; in particular they cannot dissolve any materials, hence there is no formation of ions and no conduction of electricity. For this reason very low levels of moisture content cannot be measured. But these low levels of moisture content are of no significance, for such tightly absorbed water (sometimes called 'bound' water; see Chapter 2) makes no contribution to the physical or biological properties of the material. Bound water does *not* promote decay. Thus conductance measures 'free' water, that is, water which is free to produce the effects on the material usually associated with dampness.

In practice, to measure conductance it is necessary to make electrical contact with the material under test. The conductance type moisture meter does this by means of two sharpened steel pins which are pushed into wood, or pressed firmly into or in contact with harder materials such as brick or concrete.

Capacitance
Capacitance is commonly measured between two metal plates arranged parallel and facing each other, but not touching, so that no electricity can pass from one to the other. Such a system is called a capacitor. If one plate is now charged positively and the other negatively, it requires a certain small electric charge to do so. The ratio of the electric charge to the potential difference is called the capacitance of the system, and its value depends on the area and separation of the plates, but more especially on what is between them. If it is air, the capacitance remains very small. Some substances greatly increase the capacitance; this happens although the substance between the plates does not touch either of them. Indeed, if a conductive substance were to touch both plates, they would short-circuit the capacitor and the capacitance would be effectively zero.

Water has an exceptionally high ability to increase the electrical capacitance of a pair of metal plates, about 80 times that of air. The dry matter of a brick wall, however, is only about four or five times as effective as air. Thus, if the capacitance of a pair of plates is measured, first with air between them, and then with a damp wall between them, a large part of the consequent increase in their capacitance is due to the water it contains. This is the basis of moisture measurement by capacitance. As in the case of conductance, the most tightly adsorbed water ('bound' water) is not measured.

However, in practice it is seldom, if ever, possible in buildings to reach both sides of the material under test while continuing to connect the electrodes to the measuring apparatus by the very short wires which are necessary in order to avoid electrical losses.

Therefore practical capacitance type (or 'dielectric') field instruments depend on the very small spread of the electrical field which occurs from a plate placed on one side of the material only. This is sometimes called the 'edge', or 'fringe-field' effect.

To use a capacitance meter, the sensing head is placed against the surface; it must be very close to the surface at an even distance over the whole sensing area.

How do the two types of meters compare?
Figures 3.1A, B and *C* show how the two types of instrument work in practice.

The solid and dotted lines show the range of effect and sensitivity of the instruments; they respond in greatest degree where the solid lines are passing through the material and in less degree where dotted lines are shown. Measurements below the surface can be made, as indicated in *Figure 3.1C*, by using electrodes (pins) which are insulated on their sides and make contact only at their points. For use in wood this type of electrode is supplied with a movable weight or hammer with which it can be driven in to the desired depth. For survey work in walls, floors and ceilings, somewhat longer, 'deep wall probes' are available for which holes must be drilled. The depth at which these can, in practice, be used, is limited only by the depth to which holes can be drilled.

It will be seen that a conductance type meter responds most strongly to current passing directly between the pins but in lesser degree to current passing by longer routes deeper into the material.

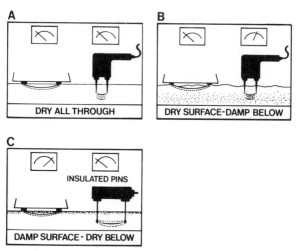

Figure 3.1 Conductance and capacitance meters

The diagrams show the range of response of the two types of electrical moisture meter. The capacitance meter, which has a flat plate electrode, is shown on the left of each diagram, and the conductance meter, which has a two-pin electrode, is on the right. Lines indicate the zones of response to water, and the little inset dials symbolize the readings given by each type of meter. Diagram C symbolizes pins with insulated shanks which can be inserted to a considerable depth in wood or, by drilling, into other building materials

The 'edge effect' of the dielectric instrument falls off very rapidly indeed away from the energized plate; therefore the solid line representing the zone of greatest sensitivity is shown only a millimetre or two away from the measuring plate, zones of lesser effect extending a short way further in.

Thus the sensitivity of the dielectric instrument depends directly on the closeness of the measuring plate to the wall.

Any inert material which prevents the measuring plate from coming very close to the wall, e.g. plastic or glass, will seriously affect the readings. Even normal surface roughness of most building materials, e.g. brickwork, prevents close contact and lowers the response of the meter to the moisture present in the material. Thus a falsely low reading can be obtained on a dielectric type meter simply owing to the unevenness

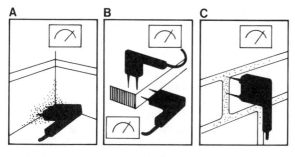

Figure 3.2 Conductance meter used in awkward places

The diagrams symbolize the usefulness of the two-pin conductance type moisture meter. Readings are being taken in important areas where a flat plate type electrode could not be used to obtain selective readings

of the surface or the fact of its being covered with an inert material. Another difficulty arises with the dielectric meter where it is not possible to place the plate flat against the surface, as in a corner or on an edge, or to take readings in limited zones such as the mortar line in brickwork.

In such situations, illustrated in *Figure 3.2, A, B* and *C*, the pins of the conductance type meter can be used without any difficulty. However, an inert material, unless it can be penetrated by the pins, will prevent a reading from being obtained in the same way as with a dielectric meter.

Conclusion
From the foregoing it can be concluded that both types of instrument respond generally to surface moisture, but the conductance type responds to the moisture at the point of contact and the measurement is un-affected by surface roughness. In addition, with the conductance type instrument it is possible, if required, to ignore surface moisture and obtain readings at any depth by using special electrodes inserted into drilled holes (*Figure 3.1C*). These electrodes are insulated except at the tip and thus make contact at any required depth.

When interpreting results it should be borne in mind that the presence of certain salts on the surface of a wall can affect meter readings. How to deal with this problem is explained in Chapter 4. Also carbon, which is present in some breeze blocks in the forms of cinders or coke, and the black colouring in some wallpapers will conduct electricity in similar fashion to a metal; so, obviously, will metal foil used as a facing on some insulating materials and sometimes as a moisture barrier. The very absurdity of maximum readings being obtained all over such a wall will at once show that the instrument is not measuring moisture.

These rather rare exceptions being borne in mind, it is a good general rule that a high reading on a conductance type electrical moisture meter indicates trouble and the need to take action.

4

Salts contamination in wall surfaces caused by rising damp

The source of rising damp is the soil or subsoil. These are always wet; without moisture in them plants could not grow. But soil consists, in the main, of decaying plant material and bacteria, moulds and soil-living animals. So soil water is not pure water, it is a dilute solution of the soluble materials to be expected from the nature of its source. Of these, nitrogen-containing salts (which eventually become nitrates), are the most characteristic, and chlorides are also generally universal. Rising damp is therefore a rising, dilute solution of various materials including nitrates and chlorides. When the water evaporates these are left behind, because they cannot evaporate. Although rising damp is a slow process, and the solution is dilute, continuance for a number of years results in quite high concentrations of certain nitrates and chlorides at the surfaces from which evaporation occurs. This is the basis for diagnosis of rising damp by analysis of wallpaper, or surface scrapings of plaster, for the presence of nitrates and chlorides. *Figure 4.1* shows a section of a wall in which rising damp had continued for 80 years. The concentrations of salts found by Building Research in the wallpaper, plaster and brickwork are given. It will be seen that the highest concentration is in the wallpaper, at the highest point to which the dampness has been rising (*see also* Chapter 5). However this is not always so as the paper may not have been in position for many years. Note

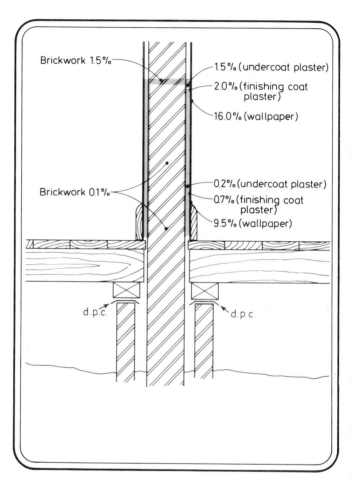

Figure 4.1 Concentration of salts in a party wall in which rising damp has persisted for 80 years. The figures show the percentage by weight of chloride plus nitrate. The shaded area is heavily contaminated

[Redrawn from Building Research Establishment Digest No. 245, revised 1969]

Box 2

Measurement of salts deposited on a wall

The analytical data given in *Figure 4.1* show that the mixture of salts deposited by evaporation from a wall affected by rising damp is strongly concentrated at the surface. Note that the wallpaper has about 13 times the concentration (by weight) found in the finishing coat of plaster. And the finishing coat has 3½ times the concentration in the base coat. Even at the upper margin of the salt contamination zone where, as usual in long-established rising damp, the salts extend more deeply into the wall, the wallpaper has eight times the concentration found in the finishing coat of plaster.

Obviously the concentration found in a sample will depend on how deeply into the wall the sample is taken. It is usually specified that the sample should be taken by 'scraping' the surface, which implies removing material to a depth certainly no greater than 1–2 millimetres. It should include the plaster finishing coat (if any), perhaps the whole of it if it is only a millimetre or so in thickness, but nothing below this. It should also include the wall covering (paper or paint) unless this is very new. If the plaster itself is also new (less than one year, say) there will be no point in taking samples anyway, because evaporation will not have had time to concentrate salts in it.

However, the sample is often not taken strictly to this specification. Therefore a quantitative (weight/weight) measure of salts concentration in a wall sample is impossible to interpret. However, *any* measurable quantity of nitrates and chlorides in a wall sample is some evidence of contamination by soil salts. A high concentration of nitrates (certainly) and chlorides (probably) shows that evaporation of soil water has continued for a long time. The depth of sampling cannot exaggerate the concentration of salts, but it can underestimate it. Therefore a semi-quantitative analysis, distinguishing between low, medium and high, on a sample volume basis, is all that is required. More precision is misleading and impossible to interpret, since the analyst has no control over the sampling.

also the relatively low concentration in the thickness of the brickwork.

Soil salts are hygroscopic, that is, they absorb water from the atmosphere (unless this is very dry) and form a solution. Hence they may keep the wall surface damp even though the rising damp which deposited them may have been cured. Characteristically, such dampness is seen by occupants of a building to vary with the weather, becoming more obvious when the atmosphere is humid.

These salts are also electrically conductive, and will give unduly high readings on an electrical type moisture meter. However, their presence is a sign of continuing trouble if the source is not cured and the contaminated plaster removed and replaced. It follows that if a meter gives a high reading, there is a condition which requires investigation even in a wall that seems to be dry.

The inner surfaces of walls in ancient buildings sometimes develop a concentration of salts not necessarily related to rising damp. These can cause high readings on the conductance type meter. This is presumably due to occasional rain penetration under exceptional circumstances over the years, which has leached these salts from the fabric of the building. Such salts are usually not hygroscopic.

Salts Detector

The Salts Detector is a Protimeter instrument for the determination of the presence of electrically conducting salts contamination on a damp surface. The instrument has a resistance measuring circuit and an analog meter display. The circuit is connected to four contact studs mounted on the surface of the instrument. A supply of small round circles of absorbent paper is provided together with a soft-surface pad on which the papers are laid.

Box 3

PROTIMETER LABORATORY
SERVICE

A guide to interpretation of results of plaster, brick or wallpaper analysis

The Protimeter analysis, being a semi-quantitative analysis, can distinguish between low (*) medium (**) and high (***) concentrations of salts. The degree of contamination by soil salts is a measure of how long the damp has been rising. A *** result of nitrates shows evaporation of soil water has continued for a long time, perhaps several decades.

How to take a sample:

(1) **Do not** take a sample...
   - (1.1) from a wall surface which has recently been disturbed by re-papering or re-plastering, for salts will not be present although damp may still be rising;
   - (1.2) from a surface which has been covered with an impervious material such as vinyl wall covering (salts will not be found in the usual place – see Box 2 – because evaporation is inhibited);
   - (1.3) from a solid floor, as the area of evaporation is too great and any concentration of salts too small.

(*cont.*)

(2) **Do** take a sample...
from a surface which has been undisturbed for a number of years just below the top line of damp. In the case of wallpaper, carefully remove a piece 50 mm square. In the case of plaster, scrape the surface not deeper than 3 mm, collecting about two teaspoonfuls.

Five points to be borne in mind when taking a sample:

(i) The diagnosis of the origin of salts should be made considering other information available, such as the moisture distribution within the building, and not the presence of salts alone.

(ii) Contamination of building materials by nitrate ions may be caused by the storage of chemicals such as fertilizers and from other sources such as urine.

(iii) Chloride ions may be present in traces in some building materials. In areas close to the sea, chlorides may be present due to sea spray or the use of unwashed sand in the construction of a building. Another source of contamination may be the re-charging of water softeners.

(iv) Efflorescent salts (containing carbonate and sulphate ions which are always present in building materials) merely indicate that moisture is evaporating from the structure. They are seldom hygroscopic and are not reported by the (Protimeter) Laboratory Service.

(v) In the vicinity of flue stacks, damp patches due to hygroscopic salts may be found. Traces of ammonia and sulphur dioxide contained in the combustion gases from solid, liquid or gas fuel burners can form salt deposits (ammonium sulphate) within the flue wall. These salt deposits are often accompanied by brown stains. They are not reported by the Protimeter Laboratory Service.

In use a paper is wetted, laid on the pressure pad, and then pressed against the electrodes. A low reading is given on the meter and this is noted. The wet paper is then pressed against the suspect (damp) wall surface by means of the pressure pad whose soft surface ensures close contact with a wall even if it is rough textured. Contact is maintained for a fixed time.

The wet paper is then pressed again on to the electrodes. Any soluble salts on the wall surface will have been absorbed by the wet paper and hence the meter will now give a higher reading than when tested with the freshly wetted paper. Such an increased reading indicates that moisture readings taken with the survey instrument on that wall are higher than the actual dampness justifies. On the other hand, if the reading given by the test paper after contact with the wall is no higher than the original, freshly wetted result, all readings given by the survey meter may be taken at their face value. This instrument thus demonstrates the presence of any electrically conducting salts, hygroscopic or non-hygroscopic, or their absence and enables a judgement to be made on the significance of meter readings and the need for stripping the plaster.

Another approach for measuring moisture in a salts-contaminated wall is to use Deep Wall Probes to measure in the thickness of the wall, at around skirting board height. As is shown in *Figure 4.1,* the salts concentrations at this point in the wall are negligible, even after many years of rising damp and will not, therefore, affect moisture meter readings adversely.

Further information on salts found in walls, and their significance, is contained in the example of a guidance leaflet provided by the Protimeter Laboratory to aid in the interpretation of the results of an analysis of wall scrapings for the presence of salts, and reproduced in *Box 3.*

DIY salts analysis

There is now available from Protimeter a DIY kit which enables the surveyor to carry out on site his own analysis of wallpaper or plaster samples for the presence of soil salts. It consists of two chemicals packed in sachets of the correct amount for each test, a bottle of distilled water, and the means for measuring the required amounts of solution and of the sample to be tested. A colour change will indicate whether or not certain nitrates and chlorides are present – and the level of contamination by nitrates (see Box 3). The test is very simple and takes just a few minutes. The kit contains sufficient chemicals for ten tests and the price is a little more than that of two laboratory analyses.

5

The sources of water causing dampness: liquid water

As stated in Chapter 1 it is convenient to classify the various forms of dampness by the source of the water which is responsible, and it was noted that a primary distinction is that between water which enters building as a liquid and water which is derived from the atmosphere. In this chapter we deal with the liquid sources; the atmosphere as a source is the subject of Chapter 6.

The sources of dampness which involve movement of water in liquid form can be classified as follows:

(1) Direct rain penetration through the structure
(2) Faulty rainwater disposal (gutters and downpipes)
(3) Faulty plumbing (water supply or disposal)
(4) Rising damp
(5) Dampness in solid floors

Rain penetration (see Figure 5.1)

The problem

Lateral penetration of rain into brickwork may be due to high porosity of the brick or to failure of the pointing, the formation of hairline cracks in rendering, or lack of adequate protection or weathering on projections outside the building. Dampness due to rain penetration is most frequently found on the south or southwesterly elevation in the UK (and on elevations

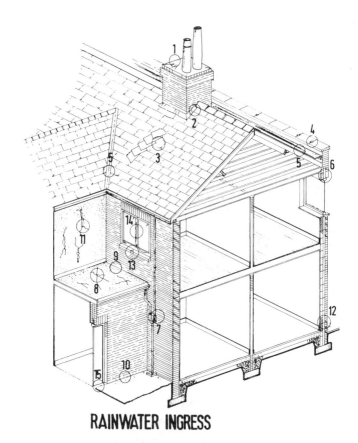

RAINWATER INGRESS

Figure 5.1 **1**, Poorly capped chimney; **2**, faulty rainwater protection (cement flaunching) where chimney stack passes through the roof (lead or zinc would be better); **3**, displaced roof tiles; **4**, faulty coping stone, without damp-proof course; **5**, defective parapet and valley gutter lining; **6**, lack of cavity tray over window head; **7**, defective rainwater pipe and hopper; **8**, cracked felt on flat roof and incorrect fall to flat roof; **9**, missing lead soaker adjacent to flat roof; **10**, rain splashback over damp-proof course; **11**, defective render; **12**, mortar droppings on ties between the two leaves of a cavity wall, transmitting penetrating rain to the inner wall; **13**, defective window sill admits water; **14**, poorly painted window frame; **15**, no threshold to prevent driving rain

facing prevailing moisture-laden winds elsewhere) and in solid, non-cavity walls. Moisture meter readings taken on the internal surface of the wall will often show local moisture zones which are adjacent to projections on the outside, or to local breakdown of the brickwork, pointing or rendering. Usually there is evidence of several high moisture zones but more general rain penetration may occur if failure of brickwork or pointing is to blame.

The main function of a cavity wall is to prevent this form of dampness. The comparatively thin outer 'leaf' of the wall, commonly only one brick width thick (4½ inches, 114 mm) is unlikely to be fully resistant to penetration. Probably it is normal for driving rain to penetrate the outer leaf to some extent. But the cavity isolates the inner leaf from this so that penetrating water runs down to damp-course level where it is usually harmless. Window frames which bridge the cavity should be protected by a tray or waterproof layer so that water is diverted harmlessly to either side, or outwards through weepholes. If mortar is spilled on to the ties which connect the inner and outer leaves of a cavity wall this may form a series of bridges conveying water to the inner leaf and this may be picked up as a series of damp spots on the wall inside (*see* p. 93). Also, mortar spilled to the bottom of the cavity may be responsible for a damp patch at skirting level inside giving some of the appearance of rising damp.

The following is quoted from British Standard 5250: 1975:

Although a simple cavity adds appreciably to the insulation of a heavyweight wall, thermal resistance can be significantly improved by placing lightweight insulating material in the cavity. This may be done either by placing material such as lightweight slabs in the cavity as the wall is built or by filling the cavity subsequently, either by pumping in foamed plastic or by blowing in mineral fibres. Cavity fill material should remain dry and should not transmit water across the cavity. Some materials and methods may be unsuitable for walls in exposed situations and in all cases it is essential that the basic construction should be free from defects that would contribute to internal dampness. Where masonry walls are to have foamed plastics or blown fibre cavity fill it is advisable to delay the application of the insulation until it has been established that the walling is not susceptible to water penetration.

Sticky polystyrene bead filling is claimed not to take water from one leaf to the next. Instead this fill permits water to run down as it does in an unfilled cavity.

The answer

A breakdown in pointing and rendering must be attended to by a competent builder. If a wall is excessively porous it may be waterproofed by application of proprietary waterproofing liquid, by spray or brush to the exterior surface. However, these materials must not be used if there are cracks in the brickwork or mortar because, by reducing the absorbancy of the wall, they increase the amount of rain which will actually penetrate such cracks. The best cure for such walls, but an expensive one, is to cover the wall, or at least its upper part, with timber or plastic cladding, or tile hanging, with an air space behind. Filled cavities must be raked out by removal of bricks where required.

Faulty rainwater disposal

On the interior walls dampness appears in local patches near the source. In the case of slow leaks the maximum moisture meter reading will give the approximate location of the source of water. The most common failure is due to blockage of downpipes at low level with the rainwater backing up to a higher point. In many properties, gutters are often hung too low with insufficient or no slope and there may be insufficient downpipes, or some may be blocked by leaves and moss.

On shallow roofs, tiles should overlap the gutter by about 50 mm (2 inches) to prevent water running behind it. Steep roofs overlap much less to prevent water being thrown out of the gutter. When houses are built with minimal eaves, so that the guttering is close to, perhaps even in contact with, the wall, the slightest error in gutter hanging will result in roof water spilling down the wall and perhaps getting behind a loose rendering.

Another source of trouble occurs when a gutter is placed actually along the top of the wall, so that there is no eaves overhang at all. While this method of construction is probably no longer used, it is not uncommon in houses 80–100 years old. When constructed, no doubt the system was effective, but the slightest leak will quickly saturate a wall. The solution is to place a flashing under the gutter and over the top of the wall.

Box gutters are often found between two lean-to roofs, behind parapet walls, on flat roofs or on boundary walls. They are usually not strong enough to support the weight of a person and are therefore frequently damaged not only because of old age but simply by someone standing in them (perhaps to carry out maintenance work, or to inspect them!). Damage cannot usually be made good in patches. To do a satisfactory repair it will be necessary to lay lining on fresh asphalt over the whole guttering (drying it thoroughly first). Box gutters, like flat roofs, are often constructed with insufficient slope. If they are supported by wood members it must be remembered that wood may bend gradually under load, and it may warp slightly as it dries out, if it was very wet when installed.

Such movements have no structural significance but they can result in the proper slope being lost so that permanent pools develop. In this event, even the most minute leak, which would not be significant in a well-drained surface, can result in a considerable amount of water entering the building.

Gutters are best examined (at some discomfort) during a period of heavy rain.

Faulty plumbing

This has to be examined carefully, including those places not easily accessible, such as in ducts and under baths. Pinpointing the damp area with a moisture meter will give a clue to a leak in an embedded pipe (*see* Chapter 7).

In hot water central heating systems joints between pipes and radiators may leak because of expansion and contraction.

Where a galvanized iron tank is used for the storage of water and distributed to fittings through copper pipework, galvanic (or electrolytic) corrosion of the zinc coating will occur where the two metals are in contact. This corrosion can be avoided by suspending a sacrificial magnesium anode in the tank thereby transposing the corrosion from the zinc to the magnesium. Sacrificial anodes last about four years.

Corrosion can also be caused by the excessive use of chemical bleaches when cleaning sinks, particularly if allowed to remain in traps for any length of time. Therefore always use plenty of clean water after using bleach.

A different, but related, problem frequently arises because, as a washbasin, bath or sink is used, a gap often develops between the fitting and the wall. This allows water to run behind the fitting possibly wetting a wooden floor so as to cause rot. To repair use a rubber based sealer which remains elastic (not cement or putty which become brittle).

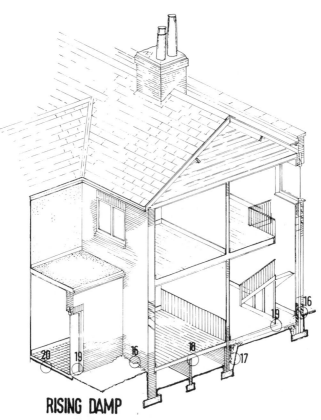

RISING DAMP

Figure 5.2 **16**, Earth or path bridging damp-proof course; **17**, earth retaining wall not tanked (i.e. no vertical damp-proof membrane) leading to a wet wall and a very high humidity in the cellar. No air brick or other ventilation to the cellar; **18**, missing damp-proof course under joists resting on sleeper wall; **19**, missing damp-proof course under floor and door frame; **20**, when a solid floor is persistently very wet, this may be due to a faulty or missing damp-proof course

Rising damp *(see Figure 5.2)*

The problem

Rising dampness results from capillary flow of water from the ground. In the absence of an adequate

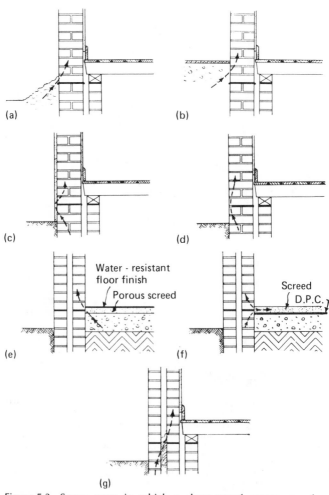

Figure 5.3 Seven ways in which a damp-proof course may be bridged, resulting in rising damp. (a) Bridging by earth; (b) bridging by path; (c) bridging by rendering; (d) bridging by mortar pointing; (e) and (f) bridging by floor screed; (g) bridging by mortar dropping in cavity

damp-proof course a damp zone extending from skirting level to about 50 cm (20 inches) or more above the skirting along the whole length of a wall may result. Where a damp-proof course (d.p.c.) exists local patches of rising dampness may result if it is broken or if it is bridged by soil in flower beds, or a stone or concrete patio is built up above it (*Figure 5.3*). After a building has been erected there may be changes in water table, perhaps due to removal of trees or new buildings interfering with normal drainage. A persistent rise in the water table up to, or even above, the d.p.c. will inevitably cause rising damp.

The rise of water in capillaries is caused by the affinity of building materials for water. All building materials are wettable; most of them are wet when the building is constructed. Therefore liquid water, when it comes into contact with a dry building material,

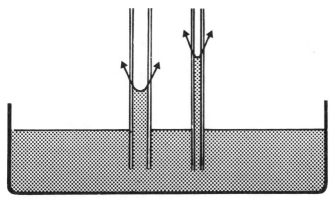

Figure 5.4 Water rising in tubes with wettable surfaces. The inner surface of a fine tube, or 'capillary', is here made of wettable material. Water tries actively to spread over a wettable surface and this produces a lifting force, indicated by the arrows. Gravity pulls the water downwards and so the surface forms a curve, supported only by the edges in contact with the wettable surface. The finer the tube, the higher the water rises. Building materials have fine pores or capillaries and are wettable. Therefore water rises in them, potentially to a great height because the pores are very fine

actively spreads over all the surfaces of the fine cracks and pores and is thus absorbed into the material, just as it is absorbed by blotting paper. This urge of water to wet any surface for which it has an affinity can be seen in a fine bore glass capillary tube when it is dipped in water. In *Figure 5.4* the arrows indicate the lifting force produced by the water trying to wet the surface. The weight of water pulling downwards pulls the surface into a curve called a meniscus – convex downwards, as shown in the figure. If tubes of different diameter are used, it will be seen that the finer the tube the higher the water will rise. The cracks and pores in brick, mortar, concrete, plaster and wood are very fine indeed, so that water could rise as a result of 'capillary attraction' to a great height. But because the cracks and pores in a wall are not continuous, the process is very slow indeed and in practice evaporation sets a limit; that is why rising damp is usually limited to ½ to 1 metre unless evaporation is prevented. If evaporation is prevented by, for example, painting the walls inside and out with an oil-based paint, or lining the walls with foil, rising damp may reach well up to first-floor level.

The answers

After any treatment for the cure of rising damp it will be necessary to remove and replace the plaster in the affected areas in order to remove the hygroscopic soil salts (see Chapter 4) which have been deposited. If these are left they will prevent successful decoration and will keep the wall damp by absorption of water from the atmosphere.

The easy cases
Rising damp is quite commonly curable, or partially so, by, for example, removal of a flower bed which was piled against the wall, or by greatly improving soil drainage combined with improved evaporation from the outside. Also remove any materials which may be

bridging an existing, perfectly adequate, damp-proof course. These would include an external rendering which may have been carried down to soil level for the sake of appearance, or inner plaster which touches a solid floor concrete slab. If very little rising damp remains, it may be cured to the satisfaction of occupants by the use of a suitable, dense, moisture resistant plaster (usually sand–cement based) in place of the original (see method 2, page 53) and the further expense of inserting a new damp-proof course (or one of the cheaper chemical substitutes) need not be incurred.

More persistent rising damp
Rising dampness which is persistent and not curable by simple means can be stopped by the insertion of a physical damp-proof course or by injection of water repellent chemicals into the brickwork.

(1) A physical d.p.c. consists of a metallic sheet or bitumen-type membrane which is inserted by cutting the wall. This is an effective but very expensive treatment.
(2) It is possible to make materials actually repellent to water, instead of attractive, by coating their surfaces with a suitable 'non-wettable' layer. The effect of this in a glass capillary tube is shown in *Figure 5.5*. The meniscus is now inverted (convex upwards) and the force which originally pulled the water up the capillary now acts in the opposite direction and pushes the water down the capillary. This is the principle of injection damp-proof courses.

In practice this method is comparatively simple: Holes are drilled with a 10 to 15 mm diameter (½ to ¾ inch) drill around the outside of the house about 150 mm (6 inches) above the level of the ground at about 100 mm (4 inch) intervals and a solution in water of a water-repellent material is

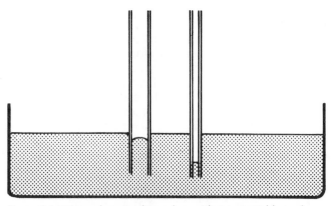

Figure 5.5 Water depressed in tubes with non-wettable surfaces. Water seeks to avoid contact with a non-wettable surface, so that the margin of the water column is pushed downwards. This is the principle of the chemical injection damp-proof courses. The solution injected lines the pores in the building material with a non-wettable (hydrophobic) substance so that water is forced out of the material instead of being drawn in

introduced using gravity bottles, or by a pressure injection process, or in the form of frozen rods. Care should be taken to ensure that the damp-proofing solution diffuses through the wall as fully as possible. Unless the whole of the wall thickness is impregnated, the result will be unsatisfactory. On completion it is usual to fill the holes using a dense 1 part cement – 3 parts sand (1:3) mix incorporating an integral waterproofing agent.

In the case of 'suspended' floors, i.e. wooden floors not in contact with the ground, underfloor ventilation should be increased as this helps to reduce rising damp by reducing the moisture content of the walls below the damp-proof course (if any).

The aim of all damp-proofing work must be to create a dry wall as soon as possible. But a wall which has been saturated by rising damp can take a year or more to dry out. Therefore the following quicker methods for producing a reasonably dry surface are suggested.

Method 1. Hiding the damp

There are various ways of concealing damp. Here are two of them:

(1) Cover the damp wall with plaster board on timber battens (the battens must be preservative-treated to prevent rotting). This method is cheap but may present difficulties around doors and windows. If the wall behind remains wet the plaster board itself may gradually become wet because of the damp air trapped behind it. The result will be development of mouldy areas, probably after some years. To prevent this the air space behind the board may be vented to the room by holes or grills at the top and bottom of the covering board.

(2) As an alternative, cover the damp wall with corrugated pitch- or bitumen-impregnated lathing. The lathing, being corrugated, forms insulating cavities on the side nearest the wall and a key for re-plastering on the other side. Before fitting the lathing, the old plaster must be removed. Because this will prevent inward evaporation from the wall, the damp will gradually rise higher until evaporation to the exterior equals the rate of rise. Therefore the outside of the treated wall should not be painted or otherwise treated to restrict evaporation.

Method 2. Replastering

Replastering will be very desirable in any event and is normally recommended as part of any curative treatment for rising damp. The skirting must be removed and inspected to see whether it has been affected by rot.

Ideally, a wall which is to be replastered should be allowed to dry out completely before the plaster is removed. During drying out, all the hygroscopic soil salts left in the wall will come to the surface and so be removed with the plaster. But usually immediate replastering is required.

Hack off the old plaster to a height of at least 400 mm (16 inches) above the uppermost limit of the damp which can be detected by a moisture meter. This will remove deposits of soil salts which have accumulated in the past but will leave some in solution in the wall. Therefore replaster with a rendering which resists movement of water and salts to the surface. There are many specifications for these and a number of proprietary 'renovating' compounds are available for this purpose. Most specialist installers of chemical damp courses specify particular recipes which must be followed precisely if their guarantees are not to be invalidated. Alternatively either of the following specifications may be used:

(1)   Replaster using a undercoat of 1:3 cement–sand to which is added an integral waterproofing compound. Follow with a plaster finishing coat.

or

(2)   Apply an undercoat of aerated 1:6 sand–cement using an air-entraining agent to produce the aeration. Follow with a plaster finishing coat. This second specification is recommended on old walls which may not be strong enough to resist the shrinkage of a dense 1:3 rendering.

In either specification washed sand must be used (in the UK it must conform to British Standard 882 Zone 2/3). Before replastering, rake out the mortar joints squarely to a depth of 10 mm to provide a good key. Always leave a gap between the plaster and a solid floor; this gap will be covered by the skirting.

Monitoring the drying out

As mentioned earlier, by the time rising damp manifests itself it has probably been rising for a long time and the wall is very wet. It follows that it must be allowed to dry out. If the interior face has received treatment which will prevent drying in that direction, e.g. replastering, or application of waterproof lathing, it is essential that moisture can escape to the outside (or to the cavity in a cavity wall). Monitoring of the drying out can be effected by the use of deep electrodes. This will require drilling some holes. Drilling through the plaster from inside will probably be objected to, but the centre of a 225 mm (9 inch), or the inner leaf of a 275 mm (11 inch) cavity wall, can easily be reached from outside and the drill holes subsequently stopped and concealed. If the wall does not dry out by the end of twelve months, failure of the damp-proof course, or of any substitute which may have been used, must be suspected.

Dampness in a covered solid floor in contact with the ground

If a floor and/or floor-screed in contact with the ground is covered with a substantially waterproof covering, the moisture content of the floor below this covering will normally rise above the air-dry condition. This is correctly shown by a conductance type moisture meter if the pins are pushed through the covering. Unless very high readings indeed are obtained this is of no significance and failure of the inserted damp-proof membrane is not indicated. The explanation of this phenomenon is as follows. When a damp-proof layer is placed on the top of a more or less porous floor (e.g. concrete, flagstones, brick, etc.) the effect is to prevent evaporation from the surface and to duplicate the existing damp-proof membrane which is within or at the base of the floor. Thus two damp-proof membranes are placed in series between the high moisture of the soil and the relatively low moisture in the air. Unless the lower membrane is perfect, the very slow movement of moisture (in vapour form) which it permits meets further resistance at the floor covering so that a water potential develops across it. That is to say, the material below the floor covering must be wetter than the air above. A frequent contributory factor is that the impervious floor covering is laid before a new screed has had time to dry sufficiently. Moisture thus sealed in will dry out only very slowly indeed.

If, instead of a waterproof floor-covering the floor was left bare or even covered with a carpet and underlay (provided these allow the passage of water vapour) or with wood boards, the resistance to water movement which these provide is negligible and the floor surface will remain perfectly 'air-dry' because of the action of the built-in damp-proof membrane. This is its function; it is neither expected nor practicable that it should do more.

Normally no action is required when dampness is detected in the screed below a waterproof floor covering. Concrete, flagstones or bricks can remain wet indefinitely without deterioration, as they do in the foundations of a building below the damp-proof course. If the moisture becomes exceptionally high, however (showing a practically saturated condition), this would indicate either residual moisture of construction or a built-in damp-proof membrane which is very inadequate and could lead to lifting and curling of linoleum or to lifting of tiles and efflorescence through joints between tiles. Worse is the case where the skirting board is in contact with the screed and may become damp enough to decay. Such a condition will be detected at once with a conductance type meter used directly on the skirting.

The instructions given with these instruments warn that as a rule any reading which indicates an above 'air-dry' condition is a cause for concern. Solid ground floors and screeds are an exception to this; in their case the warning can be disregarded, and a degree of moisture tolerated provided four conditions are met:

(1) The floors are in contact with the ground and covered with an impervious covering.
(2) No decayable material (such as wood) is in contact with the damp floor.
(3) No visible deterioration (such as tiles lifting) has been experienced over a period of several years.
(4) There is no route by which water in the concrete slab can reach the walls. This can happen if damp-proof membranes are omitted or wrongly placed, or if plaster is carried down to the base concrete.

Finally, it must be repeated that for all decayable materials such as wood (or building materials in contact with decayable materials) any moisture meter reading above the 'air-dry' level must give rise to apprehension. Its cause must be investigated and clearly understood. Unless this investigation shows that the cause is transient, any decayable materials must be isolated, preserved or otherwise removed from risk.

6

The sources of water causing dampness: water from the air

The forms of dampness which are caused by water in the atmosphere are:

(1)   Condensation.
(2)   Condensation in flues.
(3)   Dampness under suspended ground floors.

The first two of these involve deposition of liquid water from the air, but the third is the result of sustained very high relative humidity levels (*see Figure 6.1*).

Condensation

The problem (including some answers)

Water is deposited on the cooler surfaces in a building especially in winter and its presence is often first indicated by the development of moulds in the most affected areas. This is characteristic of condensation, because moulds need pure water for their growth and condensed water is pure. Unlike rising or penetrating damp it is not contaminated with soil salts or material extracted from the building itself. In severe cases the amount of water deposited may be very great, causing actual pools of water on the floor, saturated clothes in wall cupboards and decay of window and door joinery.

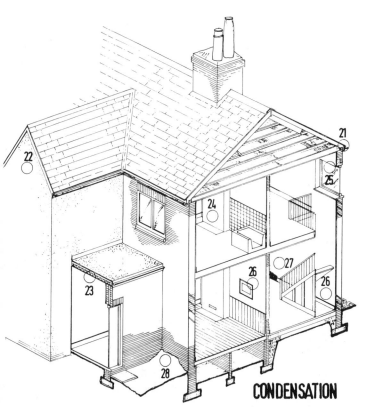

CONDENSATION

Figure 6.1 **21**, Blocked ventilation gaps to roof; **22**, no air brick to gable wall; **23**, interstitial condensation in flat roof owing to absence of vapour barrier **24**, no flue vent in blocked-up chimney breast; **25**, cold-spot condensation on solid concrete lintel; **26**, condensation behind pictures and in cupboards, due to lack of ventilation; **27**, condensation at the bottom of an external wall (looking like rising damp; *see* Chapter 7)

Occupants of affected premises often find it difficult to believe that such severe dampness can be caused by condensation alone; they frequently believe that there is a constructional defect in the building which is usually not the case. But large areas of rising or penetrating damp, or walls or floors not properly dried out after construction, can increase the amount of water in the atmosphere and may be responsible for condensation in other parts of the same premises.

Quite frequently condensation occurs predominantly at low levels where the surface of a wall is cooler, starting in the corners and eventually extending along the length of the wall. When this happens the dampness pattern may look very much like rising damp and can easily be confused with it. Two key numbers which relate to the wetness of air are concerned in condensation. These are: relative humidity (RH) and dew point. The former is a percentage and the latter is a temperature.

Water in the air

Air, like timber and building materials, always contains some water. This is in the form of water vapour, but unlike the clouds of 'steam' from a boiling kettle, it cannot be seen or felt or otherwise detected by the senses. The amount of water vapour which the air can hold is limited, but depends on temperature. The hotter the air the more water vapour it can hold; very cold air holds very little water vapour.

Saturation
When air holds the maximum water vapour possible (at its temperature) it is said to be saturated. Although this cannot be detected directly by the senses, air which is saturated, or nearly so, feels 'stuffy' whereas air which contains only a small amount of water vapour generally feels crisp and invigorating if it is cool. Warm dry air dries the nose and throat and often leads to a hard dry cough.

Relative humidity
Relative humidity (RH) is the degree of saturation, i.e. the amount of water vapour which the air contains 'relative' to the amount if would contain if saturated. This is often expressed as a percentage; saturated air is

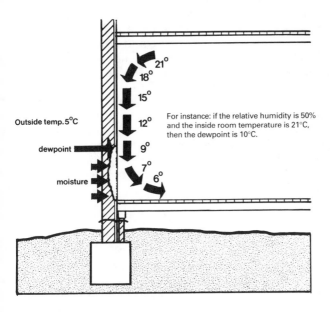

Outside temp. 5°C

21°
18°
15°
12°
9°
7°
6°

dewpoint

moisture

For instance: if the relative humidity is 50% and the inside room temperature is 21°C, then the dewpoint is 10°C.

Figure 6.2 Condensation against an outside wall giving an appearance of rising damp

In a room with a cold outside wall the temperature of which falls below the dew point temperature at low level, it is quite normal for condensation to occur predominantly on the lower parts of the walls and may be confused with rising damp. The diagram shows why this occurs. Warm air is cooled as it comes into contact with the wall and, becoming heavier as it cools, it moves downwards cooling as it does so. Eventually it reaches dew point as the temperatures on the arrows indicate, and deposits its excess of moisture. This process happens continuously so that gradually all the air in the room is involved. In practice this process is much interfered with by pictures or furniture against the walls, and 'cold bridges' may make the upper parts of the wall cooler than the rest. Nevertheless, this general tendency for condensation to predominate at lower levels, or even on the floor by the wall, can very often be observed

at 100 per cent relative humidity, whereas air containing only half what it could contain at that temperature is at 50 per cent relative humidity. But if air is warmed, the amount of water vapour it could hold at saturation is increased, so that the relative humidity becomes lower although no water has been removed from it (*see* example on p. 65).

Similarly, if air is cooled, the amount of water vapour which it can hold is reduced, so that its relative humidity is increased. If it was already saturated, the excess of water vapour which it cannot now hold must condense, either forming mist or fog (tiny drops of water in the air) or forming drops of water on walls, windows or furniture.

Dew and dew point

It is always possible to coll air sufficiently to reach the point at which water condenses to form dew, mist or frost. This temperature, that is the temperature at which a sample of air becomes saturated and produces dew or mist, is called *dew point*.

Dew point is thus the significant temperature to know whenever condensation is concerned; it depends on the amount of moisture in the air (*see* Box 4). Dew point temperature increases as the amount of moisture in the air increases. Therefore the more moisture that is produced in a room by human activity, the more likely the situation where the dew point temperature of the air will be increased to become equal to or higher than a wall surface temperature. And if, at any time, the temperature of a surface falls below the dew point temperature of the air, the air coming into contact with the surface will be cooled to its dew point temperature and will deposit the water which it can no longer hold. The cold air against the surface will sink to the floor, while fresh warm air takes its place and is in turn reduced to dew point adding more water to a cold wall (*see Figure 6.2*). A continual

Box 4

The amount of water vapour in the air

Although in the text we speak of the 'amount' of water in the air, we do not say how this is measured. The 'amount' is normally expressed as the weight of water contained in a given weight of dry air: grams of water per kilogram of dry air, for example. The ratio of the amount of water in the air to the amount of water needed to saturate that air is correctly known as the 'percentage saturation'. Strictly speaking, relative humidity (RH) is slightly different from this because it is measured as the ratio of the vapour pressure of water in the air to the vapour pressure of water in saturated air.

For most purposes this difference is not at all important. Towards the two ends of the scale, i.e. 0 per cent and 100 per cent, saturation and relative humidity are identical, but in the middle of the scale they differ slightly. For example, at 20 °C, if the dew point is 10 °C, percentage saturation is 52 per cent but relative humidity is 52.5 per cent, a negligible difference for practical purposes. But at higher temperatures the difference is greater: at 30 °C and dew point 18.8 °C, percentage saturation is 50 per cent but relative humidity is 51.1 per cent.

The term 'vapour pressure' needs some explanation. The moisture in air contributes to the total atmospheric pressure; this contribution is called the 'vapour pressure' of the water. It is commonly measured in millibars, though formerly millimetres of mercury was used. It is this vapour pressure which causes water vapour to move through building elements to regions of lower vapour pressure and determines the evaporation or condensation of water.

We prefer to use the rather imprecise term 'amount' in the text, because it is shorter and easier to understand than 'vapour pressure'. But we use the latter for all calculations.

slow circulation will thus transfer water continually to the cold wall, gradually drying the air in the room. But so long as the water is continually replaced (by drying washing, washing up or cooking, burning paraffin in an oil stove and by people breathing) the process of wetting the wall will continue even though other parts of the room may feel comfortably warm and dry.

In discussing condensation we shall from time to time refer to a 'vapour barrier'. British Standard 5250:1975 defines this as follows:

> Part of a constructional element through which water vapour cannot pass. In practice this is nearly impossible to achieve and in the text 'vapour barrier' refers to a constructional element which approximates well to the theoretical definition.

For example, vinyl wallpaper is a 'vapour barrier' (*see* p. 74).

The use of psychrometric charts

The relationship between relative humidity, dew-point and air temperature can be looked up in psychrometric charts or hygrometric tables published by HMSO or the Chartered Institution of Building Services (CIBS) or tables issued by Protimeter plc.

A psychrometric chart is reproduced by permission from British Standard 5250:1975 as *Figure 6.3*. The main curve shows the amount of water in the air when this is saturated, at each of the temperatures given along the bottom axis. The vertical axis shows this 'amount' of water on two scales; grams per kilogram of dry air, and water vapour pressure in millibars (*see* Box 4). The curved lines below the main curve represent various percentages, from 10 to 90, of the vertical axis and may be taken to represent percentages relative humidity. The sloping straight lines, which represent 'wet bulb temperature', can be ignored.

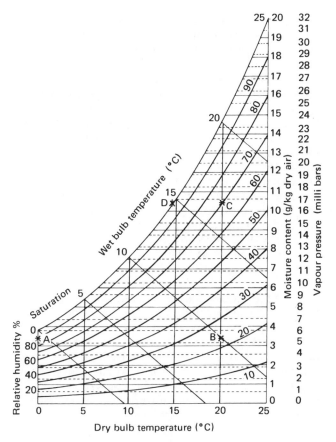

Figure 6.3 A psychrometric chart; the interrelationship between moisture contents and temperature (BS 5250:1975)

Following is an example of the use of a psychrometric chart, also taken from BS 5250:1975:

Consider point A. This represents an outdoor air condition in winter of 0°C and 90 per cent relative humidity. Point B indicates air with the same moisture content but as it is now at 20°C its relative

65

humidity has changed to approximately 23 per cent. This shows what happens to the outdoor air after it enters a building and is warmed, if no other change occurs.

Point C indicates air still at 20°C, but with moisture content raised. The increase in moisture without change in temperature means that relative humidity has risen, and the curved lines show this now to be about 70 per cent. This is what might occur when the incoming air has picked up moisture from activities within the building.

Reading horizontally to the left from C, point D indicates when saturation would occur, i.e. when the air is cooled to a temperature of about 14.5°C, which happens when the air from C comes into contact with a cool surface at that temperature. Result: 100 per cent relative humidity = condensation.

This example emphasizes that condensation in buildings is not caused by high humidities outside. So long as buildings are even slightly warmer inside than outside, air which comes in can never reach saturation indoors. The only exception is when, in early summer, a sudden change to wet warm weather may cause condensation on cold interior walls or water pipes. But this is always transient and does not cause a persistent problem.

Why condensation has become a very common cause of dampness

If all the water vapour released into the air of a home could escape somewhere, condensation would not occur. When homes were more draughty and open fires were common, this water used mainly to go up the chimneys. Now that chimneys are mostly closed and draughts prevented, condensation problems are the result (see *Figures 6.4* and *6.5*).

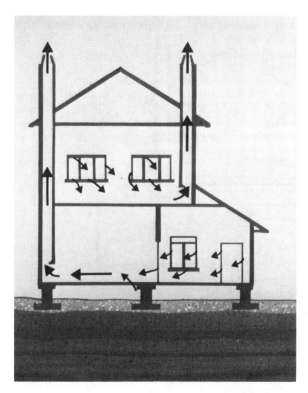

Figure 6.4 Escape of moist air in older houses

Condensation does not necessarily occur in the rooms where the water vapour is produced. A kitchen or bathroom in which vapour is produced may be warm enough to remain free from condensation except perhaps on cold, single-glazed windows, cold-water pipes and other cold surfaces. But if this water vapour is allowed to disperse through the dwelling into colder spaces such as the stair-well and unheated bedrooms, condensation will occur on the cold surfaces of those rooms, which may be remote from the source of the moisture. Soft furnishings,

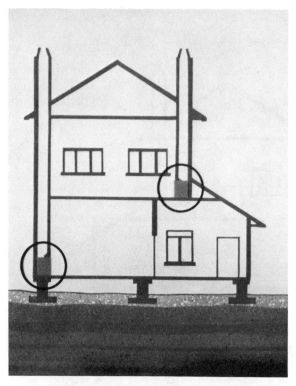

Figure 6.5 Closed chimneys and airtight windows seal the air into modern houses

including bedding, and clothing may become damp because of this, especially as some of these materials are slightly hygroscopic (taken from BRE Digest 110).

Windows

Condensation on single glazed windows is not a serious problem provided the window frames are properly painted and the condensate is wiped up regularly and not allow to soak into the wood frame and to wet the wall. In fact condensation on single glazed windows can remove quite a lot of water from the air (acting as a dehumidifier) and hence, by lowering the dew-point, may reduce condensation elsewhere. But modern man requires double glazing, which will (provided the minimum air space width between the two panes is 12 mm) reduce the risk of condensation on the glazing and if any metal frames are fitted with a thermal break this will prevent condensation on these also. However, this reduction in condensation, combined with the reduction in draughts which usually results from the use of closer fitting frames, tends to increase the risk of condensation elsewhere.

Cold areas

Another phenomenon causing condensation is cold-bridging. This means production of local cold areas on otherwise warm walls by the proximity of highly conductive building elements. For example, the inside surface of a solid concrete lintel or similar structure member may be as much as 5 °C colder than the surrounding wall surface. Such cold areas may be found in a number of places on a building; for instance where structural members such as a column or ring beam of a dense material bridge a cavity without additional thermal insulation having been provided to compensate for the loss of cavity insulation value. Another example is the bridging of a cavity with a dense floor slab which may result in condensation on the surface of the slab and possibly on the walls within the room.

Condensation in roof spaces

Water vapour will also reach roof spaces where condensation is not uncommon. In fact, the roof space may be the only place to which water vapour can escape in modern housing. With all draughts sealed and walls quite frequently covered with vinyl papers (which, as we have said, are considerably resistant to the passage of water vapour) the water vapour which is inevitably produced by occupants must escape somewhere. It will largely pass from the ground floor to upper floors and then through the ceilings to the roof space. Note that thermal insulation which is frequently applied to ceilings is fully permeable to water vapour.

Roofs with a non-absorbent lining
These types of roof include conventional constructions with a sarking felt or plastic sheeting under the tiles or slates, and constructions with metal or asbestos decks. Condensation can occur on the underside of the sheeting. This is not damaged itself, but the condensed water can then wet rafters in contact with the sheet, increasing the risk of rot. It can also drip and soak any insulation ultimately damaging the plasterboard; electrical services on the ceiling may also be wetted causing shorting. The water can also run into the eaves to soak wallhead plates and might cause corrosion to punched metal plate-fasteners in trussed rafter roofs, especially in roof members treated with a waterborne preservative based on copper–chrome–arsenate.

Problems of condensation in roof spaces may be especially severe during a thaw after large amounts of ice have built up on the sheeting as a result of condensation during a prolonged cold spell. The volume of water released may be large enough to give the impression of a major leak in the roof.

Roofs with an absorbent lining
This section is reproduced from BRE Digest 270:

> Included in these types are roofs with timber or timber-based sarking boards under the tiles or slates, or roofs with sarking boards covered with impermeable waterproofing materials. In these cases, any water condensing on the underside of the roof covering may be absorbed by the board. This may either cause moisture-swelling in materials such as chipboard or may result in rot. Under such conditions the boards may lose their strength and fall into the roof space.

If the relative humidity in the roof is persistently high, mould and mildew can grow on furniture, clothes, luggage or other materials stored in the roof. This is often the first sign of the problem and the first cause of complaint by the occupants.

Roofs with an insulated lining
The relative humidity in a roof space may be kept from rising too high by lining the roof under the tiles with an insulating material, in addition to, or instead of, insulation at ceiling level. This is sometimes done as a 'belt and braces' approach to insulation, but it is unwise and unnecessary. The insulating material may become saturated losing much of its insulating property. Also, because the roof space is kept relatively warm, air with a high dew-point can pass into the eaves, especially if these are wide, and cause condensation leading to decay of fascia and soffit boards, and rafter ends. If in spite of this, the roof surface is insulated, it is wise to provide ventilation to any enclosed eaves spaces by means of holes or grills in the soffits.

Ventilation of roof spaces

The best policy is to ensure free movement of outside air through roof spaces. This will dissipate water vapour and avoid the condensation problems described. Goods stored in the roof space will not become unduly damp; they will follow the normal outside atmosphere which, taking the year round, is far below saturation. If there is any sarking, or indeed any underlining of slates or tiles, spaces should be left at the eaves to permit free entry and exit of air. Remember that the function of a roof is to keep the rain out, like an umbrella, not to keep the warmth in (which should be done at ceiling level), still less to keep in the moisture produced in the house.

Flat roofs

The problem with flat roofs is usually one of interstitial condensation (see the following section). The external waterproof finish forms a barrier resistant to outward escape of water vapour. Prevention of this problem is aided by providing a vapour barrier located beneath sufficient thermal insulation to ensure that the under-side of the vapour barrier is kept above dew-point.

Interstitial condensation

This term refers to condensation which occurs actually within the thickness of a wall or ceiling. Any wall which is porous must allow water vapour to diffuse through it; indeed this is one of the routes by which excessive water vapour within a building normally escapes to the exterior. Very nearly always there is a temperature gradient within the wall. The inner surface of the wall is close to the temperature of the room while the outer surface is close to that of the outside temperature.

Although a wall may be porous, it imposes some resistance to the diffusion of water through it. Therefore the water vapour content of the air in the pores declines from the relatively high level which is normal in a warmed, inhabited space towards the lower level which is usual in the colder outside air. The dew-point temperature is directly related to the mass of water vapour in the air, so 'dew-point' may be substituted for 'water vapour content' in the above statement.

So we have both the actual temperature and the dew-point temperature declining through the wall. Normally, with ordinary porous building materials such as brick, the decline in actual temperature is more rapid than the decline in dew-point temperature, so that they cross somewhere within the wall and at this point condensation occurs.

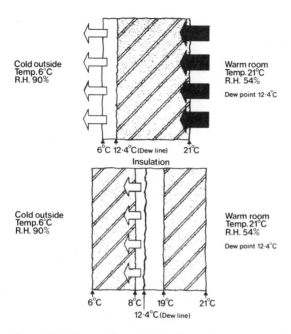

Cold outside
Temp. 6°C
R.H. 90%

Warm room
Temp. 21°C
R.H. 54%

Dew point 12·4°C

6°C 12·4°C (Dew line) 21°C

Insulation

Cold outside
Temp. 6°C
R.H. 90%

Warm room
Temp. 21°C
R.H. 54%

Dew point 12·4°C

6°C   8°C   19°C   21°C
       12·4°C (Dew line)

Figure 6.6 Two wall sections illustrating interstitial condensation. For simplicity of explanation it is assumed that water vapour passes freely through these walls so that dew-point is unchanged; in practice it would be somewhat lowered. The upper section shows a solid wall. The black arrows represent water vapour passing from the warm room. The air reaches its dew-point somewhere near the outer face of the wall and here deposits its excess of moisture. Evaporation (white arrows) continues towards the exterior and in warmer weather will normally escape harmlessly. The lower section shows the possible effect of insulating a cavity. Because of the insulation, the temperature falls most rapidly in the cavity and hence dew-point may be reached in the filling. Conditions could occur in which this filling became saturated because there is little opportunity for evaporation in drier weather. A porous filling which allows water to drain away, or a water-vapour-proof filling (acting as a vapour barrier) would not be liable to this possible hazard.

Note that an effective vapour barrier on the warm side of the wall would prevent interstitial condensation

73

This probably happens normally in the walls of most ordinary houses. If it occurs near the outer face of the wall, no trouble is caused. The excess of water evaporates quite easily at times when condensation is not occurring. Frequently, with cavity walls, interstitial condensation occurs within the cavity, because the still air presents an effective resistance to movement of heat but permits easy movement of water vapour. Such condensation normally escapes freely enough if the cavity is ventilated or may evaporate at favourable times through the outer leaf. Condensation may be more frequent, and perhaps potentially more trouble-some, when the cavity is filled with an insulating material. These conditions are illustrated in *Figure 6.6.* If the resistance of each component of a wall to water vapour penetration, and to heat movement, is known, it is possible to calculate exactly where interstitial condensation will occur. This will vary according to the weather outside and the room conditions inside so that it is usual to calculate for extreme conditions only. This is often done in the design stage for important buildings.

Vapour barriers (*see also* p. 64)

If the outer surface of a wall is covered with a water vapour barrier, heat can escape but water vapour cannot. Therefore interstitial condensation will nor-mally occur under most conditions and the wall beneath the barrier will become saturated. A cavity will avoid the worst consequences of this, but a solid wall will suffer considerably. For this reason it is an accepted rule that vapour barriers must always be placed on the warm side of any insulating layer. A vapour barrier on the inside face of a wall (for example, foil, polythene or several layers of gloss paint) will reduce the risk of interstitial condensation. For most brick or masonry walls this is probably not worth doing, but for wood frame construction it is usually considered to be essential because condensation within the wall can cause decay of the wood. However, since a total vapour barrier cannot be assured, provision should be made for moisture to escape outwards. If vapour barriers are used on all walls and ceilings, there will be no channel for escape of the water vapour which all human beings must generate in the ordinary course of their lives. If draughts are also excluded, condensation must inevitably occur.

Sources of water vapour

It is important to identify the major sources of water vapour. Some are unavoidable, but others can perhaps be eliminated or reduced and this should be done as far as possible. It is helpful when considering which sources could, perhaps, be reduced or eliminated, to know their relative importance. *Table 6.1*, reprinted with permission from British Standard 5250:1975, gives the estimated production of water vapour by a typical family of five. These sources are illustrated in *Figure 6.7*.

Figure 6.7 Internal sources of water vapour

(1) Flueless oil or gas heaters which discharge their combustion gases directly into the air inside the building can be a major contributor to condensation. This is because the burning of oil (paraffin, kerosene) or gas involves burning of hydrogen and hence production of water vapour. One litre of paraffin (kerosene) produces about 1 litre of water as water vapour (see page 84).

(2) If there is a condensation problem in the lounge, for example, and the smell of cooking gets there too, the kitchen is highly likely to be a significant source of water vapour. The obvious recommendation is to keep the kitchen door closed, as far as possible, while cooking, washing or clothes drying is in progress.

Table 6.1 shows what a large proportion of the total water vapour produced by a typical family could thereby be contained. Another effective recommendation is to install an extractor fan which will remove moisture-laden air from the kitchen.

It has been found that people are generally unwilling, or do not remember, to switch on extractor fans at the right times and hence much better results are obtained if the fan is controlled by a humidistatic switch. Also, if a fan is to be used it must be quiet and unobtrusive, and it should be provided with flaps which keep out the wind when it is not in use. Otherwise its nuisance value may be rated higher than that of condensation. If an extractor fan is not fitted, a tumble dryer should always be vented to the exterior.

(3) Large damp areas due to rising or penetrating damp, especially in warm rooms, can contribute significantly to the amount of water vapour in a building. A damp floor, if it is not covered with water-vapour-proof tiles or sheeting, can be an even more significant source. If the premises are newly constructed, time must be allowed for the

Table 6.1 Typical moisture production within a five person dwelling

Regular daily emission sources	Moisture emission per day/kg or litre
Five persons asleep for 8 h	1.5
Two persons active for 16 h	1.7
Cooking	3.0
Bathing, dish washing, etc.	1.0
Total, regular sources	7.2
Additional sources	
Washing clothes	0.5
Drying clothes	5.0
Paraffin heater (if used)	1.7
Total, additional sources	7.2
Combined total	14.4

NOTE 1. The table does not include moisture introduced or removed by ventilation.
NOTE 2. The high moisture input from clothes drying shows the importance of designing for its control.
NOTE 3. The considerable emission during cooking, which is of short duration, indicates a need for local control.
NOTE 4. The water vapour emitted by flueless oil stoves significantly increases condensation risk. (Flueless gas appliances also produce a considerable quantity of water vapour.)
[Taken from BS 5250:1975]

fabric of the building to dry properly before it is inhabited or condensation will be inevitable.

Heating

Heating of the air is in itself insufficient, although it may help to reduce condensation, since the warmed air will take longer to cool at any cold surface and give the existing ventilation more time to replace the wet air by dry air. In bad cases, however, heating can only be effective in reducing condensation if it is used long enough to raise the temperature of the cold surfaces. This seldom happens if heating is used intermittently during the evenings and mornings, as is so often the case.

In the absence of a continuous heating programme it is an advantage to place the available heat directly against the cold areas in order to warm them as efficiently as possible and to keep the air moving over the cold surfaces so that it is not in contact with them long enough to be cooled to a temperature below its dew-point. A fan heater positioned so as to blow on to the cold exterior wall would, for example, be more effective than a simple electric fire of the same wattage, but naturally would be less effective in heating the room. In the absence of any form of heating, the walls can be insulated by lining. Even a thin film of foam plastic material, such as polystyrene which is available in roll form, can improve matters to a significant extent.

Dehumidification

A dehumidifier is theoretically the ideal cure for condensation, since it removes water directly from the air and does not involve any draughts or loss of heat. Indeed, a dehumidifier warms the air by slightly more than the same amount of electricity would do if working directly through an electric heater. However, too much must not be expected of a dehumidifier. Usually the capacity of a domestic dehumidifier unit is insufficient to cure condensation in a typical house, although it will lower the general dew-point level and thus help to reduce condensation. In a single room subject to condensation a dehumidifier can be a complete cure, but as it is usually a bedroom which suffers most severely, the noise which such a machine inevitably makes requires it to be switched off at night, thereby reducing its efficacy.

In a small flat, a dehumidifier placed centrally may be effective in curing condensation completely. But this may require that the temperature within the flat is not allowed to fall too low, because the moisture removal capacity of a dehumidifier is much reduced at low temperatures. A domestic dehumidifier cannot be expected to make much contribution towards drying out large volumes of brick or concrete. Much larger (and correspondingly noisy and expensive) machines are needed to hasten drying out of new construction effectively.

The three principal objections to widespread use of domestic dehumidifiers are:

(1) High cost of installation and high running costs.
(2) Noise.
(3) Need to empty the reservoir frequently.

In the future both cost and noise can be expected to decrease, and in many circumstances it is possible to plumb the waste directly to a drain. Hence, dehumidification may become a more widely accepted treatment for condensation in the future.

Mould growth

As explained in Chapter 9, mould growth is a typical consequence of condensation. This is because intermittent periods of high moisture with intervening drier periods are especially suitable for the growth of moulds and development of the coloured spores which are principally responsible for the mould 'stain'. The distribution of mould growth can be a useful guide to the areas in which condensation is occurring, although moulds can and do develop in dampness due to some other sources of water. A mouldy area due to condensation may sometimes be found to be relatively dry when inspected with a moisture meter, because condensation has not occurred for several days. Moulds are often most severe in room corners of external walls. This is mainly because insufficient ventilation creates pockets of stagnant air in such corners. Built-in cupboards, particularly when located against external walls, suffer from the same disadvantage.

Recommendations for occupiers

British Standard 5250:1975 includes a suggested form
of explanatory leaflet which can be supplied to
occupiers of premises who are bothered by condensation. This leaflet contains a lot of useful advice and we
therefore quote it in full here.

(1) It is well known that in recent years some houses
     and flats have suffered from condensation. Walls
     and ceilings, and sometimes floors, become damp
     and sometimes discoloured and unpleasant as a
     result of mould growing on the surfaces.
(2) *Why condensation occurs.* Condensation occurs
     when warm moist air meets a cold surface. The
     risk of condensation therefore depends upon how
     moist the air is and how cold the surfaces of
     rooms are. Both of these depend to some extent
     on how a building is used.
(3) *When condensation occurs.* Condensation occurs
     usually in winter, because the building structure is
     cold and because windows are opened less and
     moist air cannot escape.
(4) *Where condensation occurs.* Condensation which
     you can see occurs often for short periods in
     bathrooms and kitchens because of the steamy
     atmosphere, and quite frequently for long periods
     in unheated bedrooms; also sometimes in cup-
     boards or corners of rooms where ventilation and
     movement of air are restricted. Besides condensa-
     tion on visible surfaces, damage can occur to
     materials which are out of sight, for example from
     condensation in roofs.
(5) *What is important.* Three things are particularly
     important:
     (a) To prevent very moist air spreading to other
          rooms from kitchens and bathrooms or from
          where clothes may be put to dry.

(b) To provide some ventilation to all rooms so that moist air can escape.

(c) To use the heating reasonably.

The following notes give advice on how you can help to prevent serious condensation in your home.

(6) Reduce moisture content of room air

(a) Good ventilation of kitchens when washing or drying clothes or cooking is essential. If there is an electric extractor fan, use it when cooking, or washing clothes, and particularly whenever the windows show any sign of misting. Leave the fan on until the misting has cleared.

(b) If there is not an extractor fan, open kitchen windows but keep the door closed as much as possible.

(c) After bathing, keep the bathroom window open, and shut the door for long enough to dry off the room.

(d) In other rooms provide some ventilation. In old houses a lot of ventilation occurs through fireplace flues and draughty windows. In modern flats and houses sufficient ventilation does not occur unless a window or ventilator is open for a reasonable time each day and for nearly all the time a room is in use. Too much ventilation in cold weather is uncomfortable and wastes heat. All that is needed is a very slightly opened window or ventilator. Where there is a choice, open the upper part, such as a top-hung window. About a 10 mm opening will usually be sufficient.

(e) Avoid the use of portable paraffin or flueless gas heaters as far as possible. Each litre of oil used produces the equivalent of about a litre of liquid water in the form of water vapour. If these heaters must be used, make sure the room they are in is well ventilated.

(f)   If condensation occurs in a room which has a gas, oil, or solid fuel heating appliance with a flue the heating installation should be checked, as the condensation may have appeared because the appliance flue has become blocked.

(g)   Do not use unventilated airing cupboards for clothes drying.

(h)   If washing is put to dry, for example in a bathroom or kitchen, open a window or turn on the extractor fan enough to ventilate the room. Do *not* leave the door open or moist air will spread to other rooms where it may cause trouble.

(7)   Provide reasonable heating.

(a)   Try to make sure that all rooms are at least partially heated. Condensation most often occurs in unheated bedrooms.

(b)   To prevent condensation the heat has to keep room surfaces reasonably warm. It takes a long time for a cold building structure to warm up, so it is better to have a small amount of heat for a long period than a lot of heat for a short time.

(c)   Houses and flats left unoccupied and unheated during the day get very cold. Whenever possible, it is best to keep heating on, even if at a low level.

(d)   In houses, the rooms above a heated living room benefit to some extent from heat rising through the floor. In bungalows and in most flats this does not happen. Some rooms are especially cold because they have a lot of outside walls or lose heat through a roof as well as walls. Such rooms are most likely to have condensation and some heating is therefore necessary. Even in a well insulated house and with reasonable ventilation it is likely to be necessary during cold weather to maintain all rooms at not less than 10°C in order to avoid condensation. When living rooms are in use their temperature should be raised to about 20°C.

(8) *Mould growth.* Any sign of mould growth is an indication of the presence of moisture and if caused by condensation gives warning that heating, structural insulation or ventilation, or all three, may require improvement.

(9) *New buildings.* New buildings often take a long time before they are fully dried out. While this is happening they need extra heat and ventilation. At least during the first winter of use many houses and flats require more heat than they will need in subsequent winters. Allowance should be made for this. It is important that wet construction should be free to dry out. In some forms of construction, especially flat roofs of concrete, final drying may only be able to take place inwards. Ceiling finishes which would prevent such drying out should not be added unless expert advice has been given that this would not matter.

(10) *Effect of increased ventilation on fuel burning appliances.* If an occupier proposes to fix an extractor fan or otherwise change the ventilation in a room containing a gas or solid fuel appliance, he should obtain advice from the installer of the appliance about the risks from toxic fumes.

Dampness associated with flues

The problem

There are two ways in which flues (chimneys) can be responsible for dampness in buildings:

(1) By conveying rain into the building.
(2) By condensation of water vapour
   (a) formed by burning of fuel (solid, liquid or gas), or
   (b) in a disused flue, from a warm domestic atmosphere.

Rainfall

Some rain will fall into an open chimney unless it is fitted with a cowl, and this will lodge at any horizontal part of the flue including the fireplace itself. If the flue is disused, this will not be evaporated and will penetrate the brickwork into inhabited areas. Also, a chimney stack which extends more than about 1 m above the roof and does not have a damp-proof course, or is poorly flashed where it passes through the roof, provides a channel through its brickwork for downward movement of rainwater into upper floors. Good protection should be provided at the top of the stack.

Condensation

Any fuel which contains hydrogen, which in practice means all except coke and anthracite, produces water as it burns, because part of the burning process is combination of hydrogen with oxygen (from the air) to form 'H_2O', the oxide of hydrogen which we know as water. In addition, solid fuels including coke and anthracite contain some water (for they are never quite dry) which is evaporated by the fire. Water from both sources is in the form of water vapour because of the heat of the fire and is therefore invisible. This gives a flue gas a high dew-point temperature. If the flue gas remains hot, all the water vapour escapes from the chimney and no harm is done. Also, if the flue gas is mixed with a large volume of room air, as happens over an old fashioned open fire, this lowers the dew-point and all the water vapour will usually escape from the chimney. With a closed stove, burning solid fuel, gas or oil, the flue gas retains its high dew-point and if the flue is cool, some of the water vapour will condense out on the walls of the flue. This will happen even if the flue is quite warm if the dew-point is very high. The worst case is burning wet wood.

Modern boilers and stoves restrict the air flow to little more than is necessary for the fire itself. Also, they are designed to extract as much heat as possible from the fire, which means that the flue gases enter the flue as cool as it is practicable to make them. It is not surprising that if a flue connected to a closed stove or boiler passes next to an exposed wall, or through cold bedrooms or loft space, it is cooled below dew-point and condensation occurs. Water deposited here, and in the exposed chimney stack above, will run back down the flue and will often penetrate the thin brickwork into inhabited rooms.

Condensed water and rainwater take up soluble salts and coloured tarry substances inside the flue and carry these through the wall where they are deposited in and on the plaster and decorations. The soluble salts are products of combustion, the most troublesome being sulphates derived from the sulphur in the fuels. All fuels contain some sulphur, but the amount varies very widely depending on the source of the fuel. Sulphates disrupt plaster and cement mortar causing it to crumble. Coloured hygroscopic deposits in walls resulting from flue condensation may cause trouble long after a flue has been disused, or even demolished.

The answer

Because so much condensation trouble occurs with closed stoves or boilers, it is necessary for flues to be lined with special lining materials. The lining prevents condensate from reaching porous brickwork and also, by providing extra insulation, keeps the flue warmer and reduces the amount of condensation.

High humidity under suspended floors

The problem

Unless subfloor ventilation of suspended ground floors is adequate, decay is very common, because the moisture content of the joists and floorboards can become very high. This is caused by the subfloor atmosphere having a persistently very high relative humidity. Wood absorbs water from air of high relative humidity and if the condition persists, the wood can become very wet indeed.

The high humidity in the subfloor space is produced by evaporation of water from the soil under the building and from the walls below the damp-proof course. This includes both the structural walls and any dwarf walls provided to support the floor. The soil is often covered with a layer of oversite concrete, but this makes little difference unless it is laid on a damp-proof membrane.

Moisture meter readings can be taken by pressing the needle electrodes through the floor covering into the floor timbers below. The pattern of dampness often follows the pattern of poor ventilation below. Typical moisture contents in wood floors moistened by subfloor humidity are 18–30 per cent; if the moisture content is 20 per cent or above, an examination of the underside of the floor is advisable. The higher figures occur when the floor is covered with a waterproof covering such as linoleum, vinyl or rubber sheet or tiles, or some types of carpet underlay.

A moisture meter with a Hammer Electrode can be used to measure moisture in joists without lifting floorboards. The insulated needles (which are 37 mm long) ensure that measurement is made only at the tips.

The answer

To prevent excessive humidity in subfloor spaces, ventilation below the floor must be increased and air vents positioned to obtain even ventilation of the whole area. Where even ventilation is difficult to achieve, the amount of water vapour generated from a soil oversite can be reduced appreciably by covering with polythene (a thick polythene sheet about 1000 gauge is effective and relatively inexpensive). However, it is difficult to do anything to reduce evaporation from walls below their damp-proof courses. Whenever there is any doubt as to the efficacy of the ventilation, the timber should be treated with a good wood preservative to prevent decay.

Dew-point (°C)   Example: ambient temperature 20 °C, RH 60%, so dew-point temperature is 12 °C. Blank spaces indicate dew-points below 0 °C.

Ambient temp (°C) \ RH (%)	20	25	30	35	40	45	50	55	60	65	70	75	80	85
1														0
2														0
3													0	1
4												0	0	1
5											0	0	1	2
6										0	1	2	3	4
7									0	1	2	3	4	5
8								0	1	2	3	4	5	6
9							0	1	2	3	4	5	6	7
10							0	1	2	4	5	6	7	8
11						0	1	2	3	5	6	7	8	9
12						0	2	3	4	6	7	8	9	9
13					0	1	3	4	5	7	8	9	10	10
14				0	1	2	4	5	6	7	9	9	11	12
15				0	2	3	4	6	7	8	10	11	12	12
16				0	2	4	6	7	8	9	11	12	13	13
17				2	3	5	6	8	9	10	11	12	13	14
18			0	2	4	6	7	9	10	11	12	13	15	15
19			1	3	5	7	8	10	11	12	13	14	15	16
20		0	2	4	6	8	9	11	12	13	14	15	16	17
21		0	3	5	7	8	10	12	13	14	15	16	17	18
22		1	4	6	8	9	11	13	14	15	16	17	18	19
23	0	2	4	7	9	11	12	13	15	16	17	18	19	20
24	0	3	5	8	10	11	13	14	16	17	18	19	20	21
25	0	4	6	8	10	12	14	15	17	18	19	20	21	22
26	1	5	6	9	11	13	15	16	18	19	20	21	22	23
27	2	5	8	10	12	14	16	17	19	20	21	22	23	24
28	3	6	9	11	13	15	17	18	20	21	22	23	24	25
29	4	7	10	12	14	16	18	19	20	22	23	24	25	26
30	5	8	11	13	15	17	19	20	21	23	24	25	26	27

7

Diagnosing the causes of dampness

How to express dampness

Dampness cannot be measured by moisture content in the range of materials met in buildings. Moisture content is the amount of water in a material divided by its weight. This means that a heavy material has a much lower moisture content than a light material which has the same amount of water in it (see p. 14).

Another problem is that, regardless of their weight, some materials can absorb much more water than others before they appear to be damp. For example, a piece of wood which is obviously dry, may have as much water in it as a similar sized piece of brick which is obviously wet.

Table 7.1 shows how misleading moisture content readings can be except for wood, even for building materials which fall under the same headings, e.g. 'bricks' or 'mortar'.

To put it another way, imagine a timber skirting in contact (and moisture equilibrium) with a brick wall complete with a plaster finish.

If the moisture content of the skirting were 12 per cent then the plaster would be at about 0.5 per cent and the brick at about 1 per cent.

Now assume the wall gets wet due to rain penetration; the skirting is at 22 per cent, the plaster between 1 and 3 per cent and the brick between 2 and 5 per cent.

Table 7.1 Moisture in various building materials

Material	Moisture content/%	Interpretation
Wood	4	Extremely dry
Some mortar	4	Dry
Some bricks	4	Damp
Other mortar	4	Damp
Other bricks	4	Wet
Plaster	4	Very wet
Wood	12	Air-dry
Brick	12	Saturated
Plaster	12	Not possible

These examples show that percent moisture (or moisture content) is not meaningful (except for wood).

What is therefore needed is a measure of moisture in materials which shows the significance of the water; whether it will allow moulds to grow, wood to decay, or decorations to be damaged, for example. Readings on a conductance type moisture meter approximate to this for they measure only the free water, disregarding the water which is bound into the material and so has no dampness effect. A high reading on the meter, after allowance for possible salts contamination, indicates a damp material regardless of its weight or its ability to

Table 7.2 Meter zones and dampness

Zone	Colour	Interpretation
1	Green	Air-dry moisture contents. Decay impossible
2	Amber or Hatched	Excess of moisture is present. This moisture cannot be attributed to high atmospheric humidity but to a definite source. If the reading in this zone persists or increases, remedial action is needed. Decay possible
3	Red	A serious moisture condition exists and immediate remedial treatment must be effected. Decay inevitable

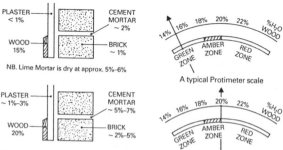

Figure 7.1 Different moisture contents of different building materials, all in moisture equilibrium. The upper example shows the air-dry (safe) condition; the lower example a wet (dangerous) condition

absorb water. Indeed, the meter will read in the green zone on all materials shown as 'dry', in the amber or hatched zone on all materials showing as 'damp' and in the red zone on the 'wet' materials. The definition of the three zones is shown in *Table 7.2* (also *Figure 2.2*).

The problem which formerly prevented a purely instrumental diagnosis of sources of damp is the effect of salts in the material. You will remember that in Chapter 4 we said that salts left by rising damp, by penetration of water from an old flue, or by leaching out of old walls over many years can cause an electrical moisture meter to over-read the moisture level. This disadvantage has now been overcome by the introduction of the Salts Detector, so that a purely instrumental diagnosis of the source of dampness is now possible.

The distribution of dampness gives clues to its source

Some case histories

A valuable clue to the source of an area of dampness is often given simply by outlining its limits and marking out degrees of dampness within these limits. This process, often known as 'pinpointing' the damp, is only possible with an electrical moisture meter,

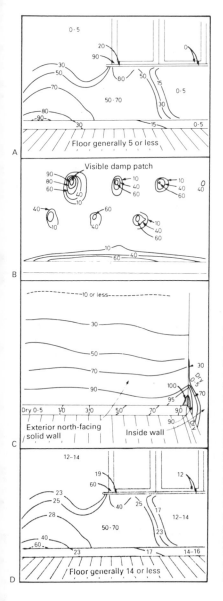

Figures A, B, C, D contain the following labels:

A: 0·5, 20, 90, 0, 30, 50, 70, 80, 50, 15, 0·5, 50-70, 30, 80-90, 30, 15, 0·5, Floor generally 5 or less

B: Visible damp patch, 90, 80, 60, 40, 10, 10, 40, 60, 10, 40, 60, 0, 40, 40, 10, 60, 40, 10, 40, 60, 10, 60, 40

C: -10 or less-, 30, 50, 70, 90, 30, 100, 0·5, 70, 95, 90, Dry 0·5, 10, 30, 50, 70, 90, 90, DRY, Exterior north-facing solid wall, Inside wall

D: 12-14, 19, 60, 12, 23, 25, 40, 25, 17, 28, 50-70, 23, 12-14, 40, 60, 23, 17, 14-16, Floor generally 14 or less

Figure 7.2 Pinpointing sources of damp. The figures quoted in A to C are taken from the reference scale of the Surveymaster moisture meter. Moisture meters of different sensitivities will give different readings. It is therefore very important when carrying out a survey to use the same type of moisture meter throughout. In D the figures are taken from the wood scale of the Protimeter Mini, Minor or, indeed, Surveymaster. They are of wood moisture equivalent (WME); see page 92

91

because this instrument is able to indicate degrees of dampness which would otherwise be undetectable. The process is fairly slow; it may take one or two hours to deal with a damp area, but the accurate diagnosis and correct treatment will usually more than reward this expenditure of time.

The technique is to take readings at regular intervals all over the suspect area, and to mark these readings accurately in position on a roughly drawn scale diagram. A tracing from an instant photograph is quick and convenient for this purpose. By joining up points of equal reading, contours of equal dampness are established.

Readings for this purpose are normally taken on a reference scale (usually 0–100) as it is only relative degrees of dampness which are required. However the wood moisture scale can be used instead ('wood moisture equivalent' or WME figures) and a wood moisture meter can be used for this purpose, provided it has a high-range circuit to enable the surveyor to distinguish between different levels of high moisture (that is, above fibre saturation level, 28%). *Figures 7.2, A, B and C* show the results of using a 0–100 scale of three such pinpointing exercises, each of which revealed much more than was apparent at first sight. *Figure 7.2A* shows a case in which incipient rising damp was suspected because of a noticeably damp patch just above the skirting board (on the left side of the diagram). The initial, limited, pinpointing exercise was extended to cover the large area shown because of unexpected indications of damp found well away from the original centre. The initial damp patch extends upwards to a considerable height without any sudden change from damp to dry such as would be expected with rising damp. This evidence that rising damp was not the cause led to a further investigation. It was traced to a plumbing leak on an upper floor producing a drip in the cavity which splashed up against the inner leaf, producing the graded dampness shown. The pinpointing exercise also brought to light a moderately damp area below the window which was shown to have its centre below the left end of the window frame. The centre being thus indicated, a fault in the sill was immediately detected and easily remedied.

This pinpointing survey (*Figure 7.2B*) was undertaken in order to clarify the source of a single small patch in the middle of a wall, the left upper patch in the diagram. A regular pattern of such patches was made evident, all undetectable by eye or touch. Also a moderately damp zone just above the skirting board was found. The pattern at once suggested that the ties connecting the two leaves of the wall were to blame. It was thought that they were either acting as thermal bridges causing local condensation, or were carrying penetrating water from the outer leaf. It was found that the latter was the true explanation. This group of ties had been very heavily loaded with spilled mortar which had also accumulated at the bottom of the cavity causing the damp area at skirting level.

The wall shown in *Figure 7.2C* appeared to be damp because of lifting of wallpaper just above the skirting board. Rising damp was suspected, but disproved by analysis of a scraping from the plaster in this region, as this showed nil nitrate. Also some mould growth was noticed which is very uncharacteristic of rising damp. The survey showed that dampness extended much further up the wall than had been suspected and declined only gradually towards dry in the upper part of the wall. This is a pattern commonly caused by condensation, when the wall is not covered with heavy furniture. The diagnosis was confirmed by wall temperature measurement. This was a solid, masonry, north-facing wall.

This survey also showed up a completely separate area of dampness which might have been noticed earlier but for the fact that the small, very wet area in the corner was partially concealed by the curling away of the wallpaper. This was found to spread for a short distance into an inside wall. Also, the skirting board was found to be decayed at the back for a considerable distance from the very wet patch, presumably because condensation was keeping it damp enough for the spread of the decay fungus, once this had been initiated by the high moisture in the corner. The cause of the wet patch was found to be an entirely unsuspected split in the rainwater downpipe, near the bottom of the wall, and on the wall side, where it could not be seen. This was saturating the wall just above the damp-proof course.

The examples show how much information can be obtained by exploiting the ability of a moisture meter to give graduated responses to degrees of dampness. *Figure 7.3* emphasizes that plumbing leaks are among the sources of damp most commonly diagnosed by pinpointing with an electrical moisture meter.

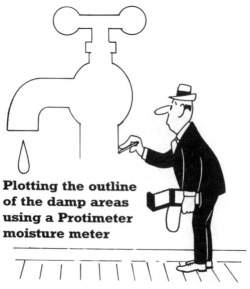

Plotting the outline of the damp areas using a Protimeter moisture meter

Figure 7.3

Diagnosis of rising damp

The term 'rising damp' is reserved for dampness caused by water rising, by capillarity, from the foundations of a building, or moving more or less horizontally from the soil where this is piled against a wall to a level near to, or above, the inhabited space inside. Inhabited spaces which are constructed below ground level are normally 'tanked', i.e. provided with a waterproof membrane on the outside, or in the thickness, of the retaining wall. If the tanking fails, soil water will enter under a certain amount of hydrostatic pressure. While the damaging effect of such entry of water is similar to that caused by rising damp, and the diagnostic features are the same, this is usually not called rising damp because it does not 'rise' and capillarity is not involved.

The two characteristic features of rising damp are:

(1) The water concerned is soil water.
(2) The water moves from below upwards and from the thickness of the wall towards its surface.

As explained in Chapter 4, soil water produces a deposit of salts, characterized by the presence of chlorides and nitrates, on the surface from which it evaporates. These salts are hygroscopic.

The diagnostic features of rising damp are therefore:

(1) Presence of chlorides and nitrates, demonstrated by analysis in a surface scraping. (For precautions to be taken, *see* Box 2 on page 35.)
(2) Damp areas becoming wetter in humid conditions and less so in dry conditions, owing to the hygroscopic nature of the salts deposit.
(3) Damp areas, which may be discoloured or stained, but are usually not mouldy, owing to the inimical effect of concentrated soil salts on moulds.

(4) Damp areas near ground level (usually within 1 metre) but possibly rising higher if the walls are treated so that evaporation is prevented.

(5) The presence of surface salts, confirmed by use of a 'Salts Detector'. (see page 36).

(6) Use of deep wall electrodes in holes drilled in the wall showing that the wall is wet internally (only possible with a conductance type meter).

(7) The skirting board, if any, will usually be wet.

(8) The upper limit of dampness as shown by an electrical conductance moisture meter will be very sharp, often changing from maximum reading to zero within 2–3 centimetres.

Diagnosis of condensation

One of the difficulties of diagnosis of condensation is the fact that it is usually an intermittent phenomenon, occurring only at certain times in each 24 hours, frequently mainly in the winter and at night. It requires a combination of a relatively high dew-point in the air and a relatively low temperature of the surfaces which are affected. These conditions may not occur at the time when a survey is conducted. Nevertheless, it is worthwhile to make measurements of dew-point and wall temperature to provide a base line from which to estimate possible differences at other times. Such differences may be estimated by questioning the occupiers. Ideally a Protimeter DampCheck should be installed in intimate contact with the suspect wall surface for at least 24 hours. This instrument reacts continuously to changes in humidity and temperature and electronically records when condensation takes place. On interrogation (by pressing a button) it will indicate green for 'no condensation' or red for 'condensation'. It can be reset to 'green' for further use by pressing a 'secret' button (secret, to prevent an uninitiated householder spoiling the test). The Damp-Check is battery-powered and is infinitely re-usable.

The diagnostic features of condensation are:

(1) Occurs on surfaces which are likely to be colder than the general temperature of the premises. These are likely to be:
   (a) behind pictures and furniture, or in cupboards against outside walls,
   (b) in bedrooms and hallways, especially in ground floor flats and bungalows where, unlike houses, they gain no heat from living rooms,
   (c) where there are 'cold bridges' such as external metal or concrete beams which come close to inner walls, and
   (d) the lower parts of walls generally.
(2) Commonly marked by mouldiness. Sometimes such areas may not be very damp, as shown by a moisture meter.
(3) Associated with very low ventilation by outside air.
(4) Associated with more than normal sources of water vapour, such as flueless heaters (gas or oil), much laundry work, including drying, within the premises, large areas of rising or penetrating damp, or even excessive numbers of occupants.
(5) Margins of damp areas are indefinite, merging gradually into dry areas.
(6) Skirting boards are normally dry, even though the wall close by is quite damp. In fact, wood generally is seldom dampened directly by condensation; any dampness which it derives as a result of condensation is usually acquired by contact with, or dripping from, other structures on which condensation has occurred.
(7) Deep wall electrodes inserted into drilled holes show that the inner thickness of the wall is drier than the damp surface.

Table 7.3 A diagnostic table for penetrating damp, rising damp and condensation

Evidence	Condensation	Rising damp	Penetrating damp
Moisture readings at the margin (especially the upper margin) of the damp areas	*Gradual* change from wet to dry	*Sharp* change from wet to dry	Usually a *sharp* change from wet to dry
Moisture readings in skirting and floor in direct contact with wall	Low readings	*High* readings	*High* reading in lower part of wall affected
Are there many mouldy patches?	Yes. They may be relatively dry at the time of survey	Very rarely	Sometimes
Is mouldiness especially noticeable behind pictures and furniture or in corners or enclosed spaces?	Yes	No	No
Soil salts, including nitrate, in wallpaper or in a scraping from the wall surface	Absent	Present[a]	Absent
Moisture readings taken at various depths in the wall using Deep Wall Probes	High at the surface, lower at depth[b]	High all through	Generally high all through. Higher towards the source and often lower towards wall surface
Sources of water vapour[c] additional to those normally met (e.g. flueless gas or oil heaters, much drying of laundry) are present	Probably one or more of these is present	These are not affected by water vapour sources, but may themselves act as a source, aggravating condensation	

[a] Rising damp may not produce a typical salts deposit if, it is derived from residual water of construction in a concrete floor slab.
[b] Dampness in the thickness of a wall may be due to interstitial condensation, but in this event the inner surface of the wall is usually dry.
[c] Further evidence that dampness is due to condensation is available if it is shown that the damp areas are likely, at times, to be colder than dew-point measured during the survey. Due allowance must be made for the possibility that conditions are somewhat different at times when survey is

Table 7.3 Summarizes the diagnosis of various kinds of damp.

How to establish when condensation is taking place

As we have shown in this book, condensation takes place on a surface when the dew point temperature of the air is higher than the surface temperature. It follows that in the diagnosis of condensation the two temperatures must be compared. There are a number of ways of doing this:

(1) The whirling hygrometer method. This instrument consists of a pair of wet and dry thermometers held in a slotted frame, which can be whirled rapidly round a handle to give a high and reasonable constant rate of airflow. The wet bulb is surrounded by a sleeve which dips into a small water reservoir. The ambient (dry) temperature and the 'wet' temperature enable the relative humidity and dew point temperature to be computed with the simple slide rule provided.

By its very method of operation, the applications of the whirling hygrometer are limited; it cannot be used, for instance, in awkward corners or in inaccessible places such as ceiling-or floor-voids. For such applications an electronic hygrometer should be used – *see below*. Also, it is very difficult to read the wet thermometer quickly enough for a correct answer, since the temperature rises the moment whirling stops. This may lead to errors.

(2) The Protimeter Surveymaster method is also based on the wet and dry principle. An absorbent paper is placed on the flat thermometer surface and fanned vigorously. Using the difference between the ambient temperature and the temperature after cooling (the 'depression'), relative humidity and dew point can be read off a table. The advantage of this method is that the

same thermometer is used for all temperature measurements.

(3) The electronic thermohygrometer method, as achieved with a Protimeter hygrometer: This gives a direct reading of ambient (air) temperature and relative humidity. Dewpoint can be read off a simple table; and an optional surface thermometer which usually plugs into the same instrument can be used to measure wall surface temperatures. If a surface is colder than the dewpoint temperature of the air, condensation will take place. The Protimeter Diagnostic now incorporates, on the back of it, a simple dew-point temperature slide-rule: the percent relative humidity measured with the instrument is set against the 'set' point, and the dew-point is read off against the ambient temperature measured with the instrument.

N.B. Measurements must be taken when all the members of the household are going about their usual chores.

A quick guide to diagnosis

| The problem | The answer |

The problem

Materials do not become visibly damp and do not feel damp to the touch until they are quite dangerously damp: it is essential when carrying out a survey not to miss any dampness

It can be difficult to make judgements about the source, the extent and severity of dampness

The answer

A conductance type moisture meter which must be able to distinguish between different levels of high moisture

---

Condensation and rising damp are often mistaken one for the other: in order not to waste a great deal of money on the wrong cure, it is essential to be able to distinguish between them

A Thermohygrometer and surface thermometer (or an instrument which combines these two functions) PLUS a DampCheck

---

Salts left by rising damp, by penetration of water from an old flue or by leaching out of old walls over many years can cause an electrical moisture meter to over-read the moisture level. It is important to establish the presence or absence of such salts

A Salts Detector and Deep Wall Probes

Differing levels of dampness must be expressed so as to make sense and to make comparisons possible

The concepts of water activity (page 25) and wood-moisture-equivalent (pages 18 and 92)

8

How to monitor the drying out of a new building structure

A great deal of water is introduced into a building during construction; some as mixing water for concrete mortar or plaster, and some as rainwater due to careless storage of building materials. Until the 1950s it was accepted practice not to decorate new houses for at least six months and then only with use of porous finishes, such as distemper or, later, emulsion paints. Wallpaper, which although in those days actually was paper and did allow the passage of water vapour (as opposed to the almost impervious vinyl type) was not applied for about a year (see also item 9 on p. 83; Recommendations to occupiers).

These are still excellent rules to follow. (Indeed vinyl wall coverings should not be applied for at least 18 months.) But modern man is in a hurry, so what can be done to produce a new dry building sooner? To dry out the building BRE Digest 163 suggests the following:

(1) Use natural ventilation by keeping windows open.
(2) Use heaters and keep windows open.
(3) Use dehumidifiers and keep windows shut (and internal doors open), applying in cold weather a little dry heat (i.e. not by means of flueless oil or gas heaters which produce water vapour as a by-product of producing heat; see Chapter 6).

It is, of course, important to monitor the drying out process. This can most conveniently be done by use of a moisture meter. To do it properly, moisture below the surface must be checked and this can be done using 'deep wall probes' (conductance type meters only).

Floor screeds

A major problem associated with the drying out of new buildings is determination of the stage at which a floor screed laid on a concrete base slab (whether laid monolithically, or separately, with or without an intermediate damp proof membrane) is sufficiently dry to permit laying of a moisture sensitive floor covering. This is often by far the slowest part of a building to dry out, the rate of drying for a 50 mm (2 inch) screed being about 1 mm per day, or ¼ inch per week. For power-floated floors it can take as long as a year. Two methods are used to monitor the drying process as recommended in BS 5325:1983.

Method 1. The Conductivity Test

A specialised instrument designed for the purpose is the Protimeter Surveymaster for Floor Screeds. This instrument is used by drilling two holes into the concrete 25 mm (1 inch) deep and 150 mm (6 inch) apart, filling these with an electrically conductive gel and inserting electrodes. The instrument is calibrated to give the average moisture content of the screed to a depth of up to 100 mm (4 inch).

Method 2. The Hygrometer Test

This method is designed to measure the equilibrium relative humidity of the screed; in practice the relative humidity of a pocket of air at the same temperature as the screed and in equilibrium with it: it is suggested that any artificial aids used to facilitate drying should be turned off four days before final readings are attempted. The hygrometer is firmly sealed to the floor and a period of not less than four hours is allowed to elapse for the entrapped air to reach moisture equilibrium with the concrete base before a reading is taken. For very thick constructions, i.e. where the damp-proof membrane is placed below the base slab, it may be necessary to leave the instrument in position for up to 72 hours before moisture equilibrium is established. WHEN A READING NOT EXCEEDING 75% RELATIVE HUMIDITY IS OBTAINED, THE SCREED IS SAID TO BE DRY ENOUGH FOR FLOORING TO BE LAID.

A Protimeter Thermohygrometer, the accuracy of which can be checked by the user, is a very much better instrument than the conventional, but unreliable, hair- or paper-hygrometer. To use it, cover a representative area of the floor with a waterproof 'tent' of polythene or foil and leave the tip of the Thermohygrometer Probe under it. Seal well with Plasticine. After 24 to 72 hours, depending on the thickness of the construction, the water evaporating from the floor into the small amount of air trapped in the tent will raise its relative humidity until it is in equilibrium. By measuring the RH it is possible to say how damp the floor is.

Method 3. The Concrete Probe

The concrete probe is designed for the measurement of moisture in structural concrete where very great precision is required. It can be used only with the Protimeter dew-point meter based on the chilled-mirror principle of dew (or RH) measurement. Readings are taken by means of drilled holes 21 mm in diameter deep within the concrete (up to 200 mm). The probe is self-sealing at any depth within the hole.

Walls

To check whether a wall is dry enough to be painted place an impervious membrane (a plastic or metal sheet about 30 cm × 30 cm) on the wall surface. Seal the edges with adhesive tape and leave in position for about 24 hours. Take a moisture reading immediately upon removing the membrane (or prick through it if it is a plastic sheet). The reading should be in the green or hatched (amber) zones of the meter. (see page 90). However, a simpler process is to check the degree of moisture within a wall by using deep electrodes (as described in Chapter 3). The surface should also be checked by using ordinary electrodes. For oil based paints the meter reading should be within the green zone. Emulsion paints are more tolerant to moisture and a reading within the green or hatched (or yellow) zones of the scale is acceptable.

Wood

Similarly, wood should be dry when painted, and it is generally considered that before application of oil based paints the moisture content of the wood should be not greater than 18 per cent for exterior work and 12 per cent for interior work. Higher moisture contents may result in blistering owing to evaporation of water from below the paint film, and consequent shrinkage of the wood. Heavy wooden members such as doorposts should be checked below the surface by using a 'hammer electrode' with insulated pins.

Timber framed buildings

This is a sound method of construction provided all the rules are observed. The most important of these are:

(1) To ensure that the wood is carefully treated with a good wood preservative to protect it from decay.

(2) To keep the wood dry during construction.

(3) To ensure proper protection against penetration of water laterally and from the ground.

(4) To ensure that the vapour barriers on the inside are not damaged.

The most likely cause of excessive moisture in the timber frame is interstitial condensation due to damage or incompleteness of the vapour barrier (*see also* page 74).

It is recommended that, at least for the first few years of the life of the house, the moisture content of the timber framing should be monitored with a moisture meter. This can be done using a Protimeter and Deep Wall Probes. Wherever necessary, small diameter holes are drilled into the frame. Deep Wall Probes with their insulated shafts are then inserted into these holes and readings taken with a Protimeter moisture meter. Between readings, the holes should be kept plugged to prevent local drying out using easily removable but tightly fitting plugs. A log should be kept of the readings; and if these show a persistently wet condition, the matter must be investigated immediately.

9

Moulds*, fungi* and wood boring insects

Fungal infections, including moulds, are initiated by the germination of air-borne spores which are always present in enormous numbers and variety in the atmosphere. Two factors are necessary for successful germination and subsequent growth of these spores:

(1) The presence of organic material to act as a nutrient source.
(2) A supply of moisture.

Organic material is present in buildings in a variety of forms; woods, wood-based sheet materials, wallpaper, textiles and horse-hair reinforced plaster and fine organic detritus; all will support fungal growth. Thus, since an organic food source is always present in buildings, the major limiting factor for fungal activity is moisture. The occurrence of moulds and fungi is therefore indicative of high levels of moisture. Although, technically, both are fungi, it is convenient to recognize two groups: moulds and decay fungi.

The actual body of a fungus, or mould, is a series of fine strands called hyphae which are often buried in the affected material and therefore invisible. However, in damp conditions the hyphae may grow out of the surface and form a mass, like cotton wool, which is

*Inspired by R. W. Berry of the Princes Risborough Laboratory of the Building Research Establishment.

called a mycelium. Fungi and moulds are therefore usually invisible or very inconspicuous until they produce spores. The wood decay fungi do this by producing a 'fruiting body', a relatively solid mass of the familiar form of a mushroom, or a bracket or simply a thick flat sheet, on which millions of spores are produced. The moulds do not make obvious fruiting bodies, they simply produce spores in clusters or bunches on the tips of special hyphae. The colour of these spores is usually the only obvious sign of mould growth. *Figure 9.1* shows the life history of a fungus or mould.

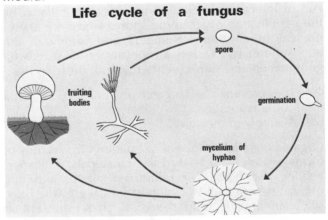

Figure 9.1 Life cycle of a fungus

Moulds

Moulds will grow on virtually any surface using fine organic contaminations as a nutrient source. Germination of their spores requires a relative humidity well in excess of 90 per cent; in all probability a film of free water is required, at least briefly. But once germination has occurred, growth will continue at relative humidities as low as 70 per cent. With such minimal moisture requirements, moulds are ideally suited to exploit even the often marginal moisture conditions arising in buildings prone to condensation. Mould growth may therefore indicate periodic, superficial wetting due to condensation, which is why mouldy patches may give relatively low moisture readings at the time of survey. Such areas may only be damp at particular times of the day or week. However, mouldiness should always be further investigated since it may also indicate a high moisture content throughout the underlying wood or brickwork, either as a result of persistent condensation or possibly some other form of water ingress, usually in vapour form (e.g. damp in a dry lining or panelling due to persistent high moisture in the wall behind). Plumbing and roof leaks are also prone to encourage development of moulds as these are reasonably pure water. It should be noted that mouldy 'stains' are not usually permanent stains; they are due to the coloured spores which the moulds produce when conditions become adverse, usually by partial drying of the surface. These 'stains' can often be brushed away when the source of damp has been removed, but it is sensible to follow this up with a fungicidal wash to delay their reappearance in case dampness temporarily returns.

Wood decay fungi

In certain circumstances some of the fungi which normally decay fallen trees are able to colonize wood in buildings. After a year or so these fungi produce large fruiting bodies which are useful in identification. Decay fungi are all limited in their activities to wood in which the moisture content exceeds 20 per cent. Moisture contents of wood in most buildings vary between 10 per cent in well heated and ventilated rooms, to levels approaching 18 to 20 per cent in areas such as roof voids and suspended wooden ground floors, particularly in winter. It can thus be seen that, in many buildings, the margin of safety from conditions suitable for development of decay fungi is comparatively small. The intrusion of some additional moisture into the fabric of a building, possibly quite minor in itself, may be all that is needed to tip the balance and to initiate the onset of fungal decay. Moisture may come from roof leakages, rising damp, condensation, penetrating damp or plumbing leaks, but without it no decay can begin.

One fungus worthy of special mention is the dry rot fungus, *Serpula lacrymans* (formerly known as *Merulius lacrymans*). Despite the implications of the name, 'dry rot', it is still limited to growth in wood with a moisture content of at least 20 per cent. Its special feature is its ability to transport water through

Figure 9.2 Dry rot fruit body at the top of a doorway. It produces millions of air-borne spores 0.01 mm long which will blow about the house in search for damp timber. When you see one of these the fungus has been active for at least a year.

well-developed conducting strands and so, to a limited extent, moisten wood which would otherwise have been too dry to support fungal decay.

Another important feature of *Serpula lacrymans* is its ability to pass through lime mortars and plasters in its search for food. Claims for the water conducting

Figure 9.3 Extensive dry rot attack (*Serpula lacrymans*) in a cellar with extensive mycelium growth. *See also* the *Coniophora puteana* (wet rot fungus) strands in the right hand corner.

Figure 9.4 Close up of *C. puteana* as shown in *Figure 9.3*

abilities of the fungus have, however, been greatly exaggerated since it is capable only of spread through areas of high moisture content or high relative humidity. Nevertheless, these adaptations to the building environment make it one of the fungi causing most rapid decay and probably the most difficult to eradicate.

Other decay fungi will always cease their activities as soon as the source of moisture is removed. But the dry

Figure 9.5 Upper diagram: Typical cubical breakdown of timber attacked by *Serpula lacrymans*. Lower diagram: Cubical breakdown of timber attacked by *C. cerebella* (the cellar fungus, wet rot). The cracking is virtually indistinguishable from that of the *Serpula* attack).

rot fungus is less easily stopped in this way because it may have spread so far within a building.

113

To sum up

Spores of moulds and spores of wood decaying fungi are commonly present in the atmosphere. They will cause fungal infections in the presence of high levels of moisture. Mould growth development is likely where surfaces inside buildings become intermittently moist and where relative humidities sometimes exceed 90 per cent. Moulds require 'clean' water, that is, not water contaminated by chlorides and nitrates such as are found in water from the ground. The presence of moulds is usually evidence of condensation.

Wood decay fungi occur when wood is at a moisture content of over 20 per cent, a level of moisture which may not be uncommon in winter in roofs and under suspended floors.

Figure 9.6 The Elf Cup fungus *Peziza spp.* growing on damp masonry. This fungus will not decay wood, but it is an indication that dangerously damp conditions exist and that rapid drying out is required if serious outbreaks of dry or wet rot are to be avoided.

How to eradicate rot

(1) With your moisture meter find the source of moisture and remedy it. Trace out the limits of the dampness in the building, working outwards from the attack which has been seen. Remember that the fungus can grow in three dimensions and that the dry rot fungus can grow right through massive walls (if these are damp).

(2) Expose (if at all possible) all faces of all wood in the area of attack.

(3) Cut out all structurally weakened wood in these areas; in case of dry rot go two feet beyond any affected wood, and replace by new wood pre-treated with a good wood preservative. Preferably use 'red wood' (Scots Pine, *Pinus sylvesris*) as it is easily penetrated by the preservative; 'white wood' (Spruce) is very difficult to preserve.

(4) Treat all remaining wood with a wood preservative. This may be brushed on liberally in order to saturate the wood until it will take up no more preservative.

(5) In case of a true dry rot (*Serpula lacrymans*) outbreak, set up physical and chemical barriers in order to isolate all wood in the area of treatment from damp brickwood. To set up such chemical barriers all reveals and walls near to wood should be given an extra soaking with a water-based fungicide which diffuses into damp masonry and destroys active hyphae. (This is better than the old-fashioned method of using a blow lamp, which process, apart from creating a very real fire hazard, is ineffective, offers only a surface treatment and has no residual effect.)

Wood boring insects

It is sometimes said that dampness does not influence attack by wood boring insects one way or the other.

This is not so. Many of the most common wood boring insects will only attack damp wood, or are much encouraged by dampness. Some types of wood boring insect have a preference for wood which is already partly pre-digested by fungal attack. The Death Watch beetle (*Xestobium rufovillosum*) (*Figure 9.7*), for example, is one which will attack only hardwoods, especially oak, which have been affected by a fungus. For this reason it is often confined to the inner parts of heavy timbers which do not rapidly dry out in dry periods. Its presence in old buildings is often evidence of moisture ingress.

Figure 9.7 Death watch beetle (*Xestobium rutovillosum*)

The presence of wood boring weevils [*Euophryum* (*Figure 9.8*) or *Pentarthrum* species] happens only where there has been attack by a fungus, usually the Cellar Fungus (*Coniophora cerebella*). This is one of the few insects whose adults, as well as the larvae (grubs) eat the decaying wood.

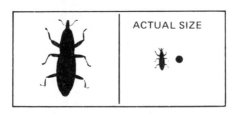

Figure 9.8   Wood boring weevil (*Euophryum confine*)

By far the most widespread wood destroying insect in Britain, and world wide in temperate climates, is the common furniture beetle (*Anobium punctatum*) (*Figure 9.9*). This insect does not need a fungus to pre-digest wood. It can therefore attack dry wood, but its life cycle is much shorter if the wood is damp. The normal life cycle from egg to adult beetle, through the larva (grub) stage (which does all the damage as it tunnels through wood, reducing it to a much weakened condition) is about three years in dry wood.

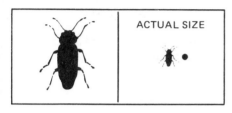

ACTUAL SIZE

Figure 9.9 Common furniture beetle (*Anobium punctatum*)

After this time a pupa is formed, then an adult beetle which bites its way out leaving the well known tell-tale exit hole. But in persistently damp wood the life cycle can be as short as one year. This reduction in the life cycle period to one third as a result of damp is of much more than academic interest because it means that the rate of increase is about nine times greater in damp than in dry wood. In practical terms, this makes the difference between a few negligible holes, somewhere in a building, and a serious infestation, which can produce major damage. It should be noted that the sapwood only is attacked unless wood-rot is present when infestation may be found in the heartwood also.

It has frequently been observed that an outbreak of 'woodworm' (nearly always furniture beetle) ceases to be serious when central heating is introduced and the woodwork of a house becomes dry for the first time. Indeed at very low levels of moisture – below 10% –

which are not uncommon in centrally heated buildings, infestation by all species is not possible. Wood boring insects, therefore, provide one more reason why all sources of damp should be removed.

Other insects

Two additional insects, termites and carpenter ants, deserve mention (according to R Bruce Hoadley, *Understanding Wood*, Taunton Press Inc., Connecticut, 1980, reprinted 1981) in view of the devastating damage they cause in stored lumber and wood structures (in the USA). Both insects are social, and their colonies have a preference for moist wood. For both, the key to prevention is keeping the wood isolated from the ground and keeping it dry.

Appendix: For the practical and concerned householder

How to survey your house for dampness

The method described can be carried out using any type of moisture meter which will give a graduated response according to the degree of dampness. Most meters will have either or both of a 0–100 scale of dampness, and a colour code (green, hatched or amber and red). This is in the form of a coloured line on the dial, or two or three coloured lights. Alternatively, a wood moisture meter will have a wood moisture content scale and also the colour code, shown by three bands of colour.

In the method described only the colour code is referred to, since it applies to all Protimeter instruments. Greater precision is, of course, given by a 0–100 scale, or a wood moisture scale of the more sophisticated instruments, but the colour code alone is quite sufficient for the basic survey.

Start your survey on the ground floor. Do one room or the entrance hall at a time. In each room proceed as follows:

(1) Make a rough plan; work round the skirting and also the floor, close to the wall; give special attention to doorways, especially the foot of each door post.

Mark in red on your plan any point on skirting

or floor which gives a red zone reading (give also the numerical reading or wood moisture reading, if your meter has this on its dial) and in yellow any amber (or hatched) code reading.

(2) Take each wall in turn. Draw a rough diagram for each wall marking in any windows, doors, fireplaces or other features. Work along the wall above the skirting, up each corner, including corners by chimney breasts or buttresses, and mark yellow and red (or numerical readings) on your diagram. Then work up the wood of each door frame and across the top. Do the same for each window frame, giving special attention to the bottom end of each upright member and the point where it joins a transverse member. Throughout, confine your measurements to wood, brick, stone, plaster or wallpaper. Metal does not absorb water and will always give a maximum reading.

(3) Try a few readings in outer walls away from corners or window frames. It is unusual to find a damp patch in this situation (except as an extension from damp already observed above the skirting board), but there may be locally cold areas ('thermal bridges') causing local condensation, or perhaps leakage from an external rainwater pipe or penetration of rain caused by a ledge on the outside wall.

(4) Dampness can sometimes be found behind pictures and furniture, usually caused by condensation. Such sheltered parts of a wall often remain cooler than the rest of a room because warm air and radiant heat do not reach them.

(5) Always give special attention to built-in cupboards and spaces under the stairs. Here, air circulation is much reduced and the tendency is to be cooler than the rest of the house. Dampness here may be due to condensation, or it may come from outside (rising damp or leakage) and fail to evaporate owing to poor air circulation.

(6) Test the ceiling in a few places if you have any reason to expect trouble. This will be commonest under a flat roof (leakage or condensation) or under a bathroom where there can a long-standing unsuspected slight leak.

(7) If the ground floor is a 'suspended' floor, i.e. raised above ground on joists, leaving a space above the ground, it will be advantageous if you can lift one or two boards in order to test the joists. If these give high (red zone) readings, it will probably be due to inadequate underfloor ventilation. This is a dangerous condition which can lead to rot and collapse of the floor. The underfloor ventilators (air bricks or grills) must be kept clear for free air circulation.

(8) If the ground floor is solid, i.e. concrete, stone or brick, laid directly on the soil (normally with a damp-proof membrane on the soil, or in the thickness of the floor) test this through the floor coverings. If the floor covering is waterproof (vinyl or rubber tiles, for example) you are quite likely to find dampness below it. Do not be alarmed, this is quite a normal condition. Read Chapter 5, p. 55, on this subject.

(9) Repeat the survey on all floors (except items 7–8).

(10) Enter the roof space and test as much of the woodwork as is accessible. Dampness here may be due to condensation if the roof space is not well ventilated, or to unsuspected leaks between tiles or slates.

Obviously all this will take more than two hours first time. But thereafter with sketches already done and the danger spots identified, the repeat survey becomes a rapid process.

Diagnosis

Correct diagnosis of dampness must come before effective cure, therefore the cause of every damp spot or area should be discovered. If a damp spot persists throughout the year it will be essential to find its cause so that it can be cured before serious damage can occur. But temporary damp spots, which are present at one survey but gone at the next, are relatively unimportant.

The source of many damp patches will be obvious from their position. Does it relate to a water pipe, interior drainage pipe or external rainwater pipe? Is there a leaky gutter or roof? Does condensation develop regularly in winter, for example on windows? Is earth piled against the wall outside? Is there a possibility of regular spillage? Questions like these can often be answered easily by the householder who knows his house intimately, especially if the dampness is in small patches.

If the dampness is in a more generalized area you should read Chapter 7, especially the diagnostic table on page 98. It will probably be found very well worth while to conduct a full 'pinpointing' investigation as described in Chapter 7, pp. 90–94.

Postscript

My friend and co-author, Professor Tom Oxley, died in 1983 shortly before this book was published for the first time. As a former director of the government Forest Products Research Laboratory at Princes Risborough (now a part of the Building Research Establishment) he was one of the best informed men of his time in the field of dampness in buildings.

He wholeheartedly supported me in my condemnation of expressing such dampness in terms of percentage moisture content. The first sentence of Chapter 7 is his: 'Dampness cannot be measured by moisture content in the range of materials met in buildings.'

He would be astounded and upset if he knew that many years after publishing these words the practice still persists. The reason that it does is explained, by those who practise it, by what they see as the danger of contamination of building materials by salts which affect moisture meter readings.

Oxley was able to support my point of view because he was convinced that the incidence of rising damp was much exaggerated by the experts, particularly remedial treatment firms. As rising damp was the main source of the offending salts so, too, in his view, the incidence of salts contamination was much exaggerated.

He was glad when we produced the Protimeter Salts Detector (page 36); for with it he was able frequently to prove his point, and to show that a high reading on a moisture meter was, more often than not, evidence of a major moisture problem.

Recently I came across a letter dated 6 August 1980 which Oxley had written, while a Visiting Professor of the Biodeterioration Centre at the University of Aston in Birmingham, to his former colleague and lately head of mycology at FPRL, Dr W. P. K. Findlay. It is a long letter dealing with several aspects of moisture in buildings. This is what Oxley said on the subject of salts (bearing in mind that this was written before the advent of the Salts Detector instrument):

'Answering your question, that is: how can one distinguish with a Protimeter (moisture meter) between high electrical conductivity caused by moisture and that caused by electrolytes (such as soil salts), I can only say that clearly one cannot. But the question is not entirely valid because it is not a true antithesis; both factors (water and electrolytes) are working to give a reading.

At the extreme, neither pure water nor pure dry electrolytes will conduct electricity, and therefore neither will give a Protimeter reading. Fortunately there are always some electrolytes present or you would never get any Protimeter readings at all. The validity of the electrical conductance measurements as an indication of the presence of water depends on the water being the principal limiting factor. If you add a lot more electrolytes (as in CCA or boron treated wood) you do increase the Protimeter readings because the minor limiting factor, the supply of electrolytes, becomes unlimited. The increase is not enormous; correction factors for the example of salts in wood are published.

The higher-than-normal readings obtained with a Protimeter on a wall whose surface is contaminated with salts is to only a small extent due to their effect as electrolytes; the greater effect is due to their hygroscopicity increasing the amount of water present. If the operator goes to the bother of drilling small deep holes in a wall and using deep wall probes, which have insulated shanks and make contact only at the tips, the concentration of salts is very low indeed and hygroscopicity cannot be working in the absence of air. Therefore, a high reading with such deep probes means that there is a lot of water still in the wall.

The statement that Protimeter readings are "misleading" in walls which have been subject to rising damp does not apply with deep probes. How far are the readings "misleading" when they are taken on surfaces? Certainly the readings will be higher in the presence of salts than they should be for the amount of water present *from within the surface,* but this is at least partially a correct reflection of the excess water which comes *from the air* because of hygroscopicity. As I see it the readings are misleading only if they are interpreted as a firm indication that rising damp is continuing. Unfortunately the claim that Protimeter readings are "misleading" in walls which have been subject to rising damp is widely used by unsuccessful damp-proofers to discount the evidence which they provide. The unscrupulous frequently use drilled samples tested in a carbide-type tester [page 23] to show that there is a "low" moisture content, perhaps only 2 or 3 per cent by weight, which looks dry to the occupier. In most mortars this would be dry; in brick it is definitely on the wet side.

The perfect fiddle used by the unscrupulous is to take a carbide-type tester reading in the mortar before treatment (to show how wet it is) and in the brick after treatment (to show how much it has dried). In the same wall, the Protimeter would give substantially the same reading in brick and mortar.'

These last few sentences sum up the reason why percentage moisture readings in building materials other than wood are not only meaningless, they may indeed be against the public interest. And what applies to percent moisture content equally applies to percent saturation since this, too, does not tell you how wet (in terms of risk) a material is.

It would be foolish to carry out a survey of a house without a moisture meter (for dampness can be hazardous long before it can be detected with the human senses of sight, touch or smell). But it would be no less foolish to rely on a moisture meter on its own for a complete dampness diagnosis. A comprehensive diagnostic kit is now available – and coming as it does from Marlow, the home of the Compleat Angler Hotel, it is inevitably called the 'Compleat' Kit. How Oxley would have loved it...

E.G.G.

Index